CELESTIAL MECHANICS

MATHEMATICS LECTURE NOTE SERIES

CELESTIAL MECHANICS

PART I

SHLOMO STERNBERG

Harvard University

W. A. BENJAMIN, INC.

New York   1969   Amsterdam

CELESTIAL MECHANICS Part I

Library of Congress Catalog Card Number 69-18393
Manufactured in the United States of America
12345 M32109

*The manuscript was put into production on March 25, 1968;
this volume was published January 1, 1969*

W. A. BENJAMIN, INC.
New York, New York 10016

In memory of Aurel Wintner

A Note from the Publisher

This volume was printed directly from a typescript prepared by the author, who takes full responsibility for its content and appearance. The Publisher has not performed his usual functions of reviewing, editing, typesetting, and proofreading the material prior to publication.

The Publisher fully endorses this informal and quick method of publishing lecture notes at a moderate price, and he wishes to thank the author for preparing the material for publication.

PREFACE

Due to the length of the manuscript these notes are appearing in two parts. The first part is a historical survey. With the exception of the Appendix to Chapter I and the last section of Chapter II, it can be read with little mathematical preparation beyond a first course in calculus.

Although a careful attempt has been made to eliminate error and some sections were rewritten many times, there undoubtably remain many errors - typographical, astronomical, historical or mathematical. The author would greatly appreciate being informed of any of these errors and any suggested improvements for possible future revision.

The research of this work was supported by National Science Foundation Grant GP-6585.

vii

TABLE OF CONTENTS

INTRODUCTION

These are notes in a course in celestial mechanics given at Harvard during the spring of 1968. The main purpose of the course was to study recent developments in the subject, principally the ideas inaugurated by Kolmogoroff and developed by Arnold and Moser. The first portion of the course was devoted to a very sketchy survey of the history of the subject emphasizing its influence on the development of various branches of mathematics.

Chapter I is concerned with ancient and medieval astronomy - that is, astronomy up to the time of Kepler. In a sense, this subject can be described as empirical Fourier analysis together with a geometro-physical interpretation of the various terms. In most popular accounts, the contributions of the Babylonians and the Greeks are consistently underestimated. However, the fact remains, for instance, that the Babylonians, in addition to having accumulated impressive observational data, based their tables and predictions on a form of Fourier analysis (using what are now known as "spline functions" instead of trigonometrical series). The authorative source for any serious study of this subject is the work of Neugebauer, in particular his "Exact Sciences in Antiquity". I have also found the "History of Astronomy" by Pannekoek very readable and informative.

Unfortunately, I am conversant with neither Greek or Latin, nor can I read cunerform. I therefore am not reporting first hand from any ancient text. However, the classical Jewish education contains (and religious obligation demands) the study of all branches of the Law . This includes the laws of the calendar and the sanctification of the new moon. I therefore have some familiarity with the two classical medieval Hebrew texts, the Sefer Ha'ibur of R. Abraham, b. Hiyah and that section of Maimonides' Code dealing with the sanctification of the new moon . I have therefore included some discussion of the astronomical portion of Maimonides' code. (I hope to publish some comments on the legal aspects elsewhere.) For the benefit of the non Hebrew reading audience, the discussion and figures follow Neugebauer's astronomical commentary rather than the standard Hebrew commentaries. A very good English translation by Gandz[G 1] (together Neugebauer's commentary) is available from the Yale University Press.

In the appendix to Chapter I, I try to show how the ancient idea of mean motion and "almost periodicity" has developed into some of the important branches of modern mathematics such as ergodic theory and group representations. Included are a discussion of Weyl's ingenious application of the Kroneker-Weyl theorem to solve the Lagrange problem of mean motion, the ergodic theorem, the Peter-Weyl theorem and Bohr's theorem on almost periodic functions. All proofs are included.

The main purpose of Chapter II is to show how number theoretical considerations (known as "small divisors") entered into celestial mechanics. Also included is a very sketchy development of Hill's lunar theory which is, in a sense, the origin, via Poincaré, of most modern work on the qualitative theory of differential equations, a good bit of functional analysis, and a substantial portion of algebraic topology. Hill's work is a remarkable combination of mathematical daring and numerical computation and precision. The complications in the moon's motion are seen in the article by Brown (who completed Hill's work) in the

Encyclopedia Americana, 1949 Edition, volume 19, page 425

of the moon's motion . We quote from it:

"In order to give an idea of its complicated motions,

a model, on a scale of 1:125,000,000, can be constructed

as follows: The first part, to represent the motion of the

earth, consists of a rail on raised supports (which are

movable) in the shape of an ellipse whose semi-axes are

respectively 390 feet 10 inches and 390 feet 3 inches long an

and whose plane we shall for convenience take to be hori-

zontal. The earth is represented by a carriage moving on

this rail and the sun by a ball placed on the longest diameter,

distance 6 feet 7 inches from the centre of the ellipse. A

straight bar two feet long is attached to the carriage by a

ball and socket joint at a point 2/3 inch from the centre of

the bar. The bar forms the longest diameter of an elastic

tube in the shape of an ellipse and the tube is so attached

to the bar that it may change its size and shape slightly.

The plane of the tube is to be inclined about 5° to the

horizontal. A bead sliding freely within the tube represents

the moon.

Motion. Now let the carriage run along the rail not

quite uniformly but so that its angular velocity about the

ball representing the sun varies inversely as the square of

its distance from the ball, and it makes it a complete circut once in 365 1/4 days. At the same time the plane of the tube attached to the carriage is to turn round slowly in the opposite direction, so that the horizontal line in it (that formed by the intersection of the plane of the tube with the horizontal plane and called the line of nodes) will describe a circuit once in 18 2/3 years; the inclination of the tube is to oscillate not more than 5' on either side of the mean inclination of 5° 8' to the horizontal. With these motions, the bar carrying the tube is to slowly turn round in its own plane in a forward direction so as to complete a circuit once in about nine years, and the tube is to slightly change its shape and size to and fro as it moves, finally, the bead representing the moon moves according to the same law and in the same direction as the carriage, that is, not quite uniformly, but so that its angular velocity about the joint varies inversely as the square of its distance from the joint; its circuit in the tube is completed once in 27.5546 days - the anomalistic period. These various motions, complicated as they are, only give a general idea of the way in which the moon moves, but the model is sufficient to explain most of the phenomena connected with the moon's motion. All the parts are in a state of oscillation about their average positions, the periods varying from a few

days to many thousand years. Even the plane, size, shape
and position of the rail are not quite constant but vary
slowly from year to year. "

In a certain sense, the work of the classical
astronomers in perturbation theory is no longer relevant
for predicting the motion of the planets. The advent of
radar astronomy (and to a certain extent the space probes)
has given observational accuracy for surpassing anything
known heretofore. This information is incorporated into
the equations of motion which are numerically integrated
on high speed computers and the various numerical constants
are continually readjusted to fit the data. This work,
mainly being done by Ash at Lincoln Laboratories,allows
computer prediction of planetary positions far more accurate
(by brute computation) than anything provided by classical
perturbation theory. In a very real sense, one of the most
exhalted of human endeavors, going back to the priests of
Babylon and before, has been taken over by the machine.

The qualitative predictions of the perturbation theory
and , even more the precise results of the Kolmogoroff
school presented in Chapter III are still of importance - in
plasma physics. Here (in contrast to the planets) the time
scale of the events is the same as that of the computer so

that the computer cannot keep ahead of the events, and a
qualitative prediction is necessary. We refrain from
discussing this subject, however.

The beginning of Chapter II begins with a brief
discussion of a portion of Kepler's work. I concur with
Koestler's view that Kepler's work represents the turning
point in the development of physics and astronomy and
therefore, if such a statement makes any sense at all, he
represents one of the most important figures in all of
history. Pedantic criticism notwithstanding, I recommend
Koestler's biography of Kepler, "The Watershed".

The last section of Chapter II deals with some of the
astronomical implications of general relativity. It is only
recently that there have been quantitative experimental
tests of general relativity. Presumably, by the time of
publication of these notes there will be conclusive results
from the time delay test of Shapiro, Ash, et. al. There are
many books on general relativity, some of which are good.
In all those that I am aquainted with, it takes a substantial
amount of reading to get to the astronomical implications
of the theory. I avoid this by the simple device of starting
in the middle. This section assumes a modest aquaintance
with differential geometry and omits some simple but messy

curvature calculations. Otherwise it is self contained.

Chapters III and IV represent the heart of the course and can be read in either order. Chapter III is an exposition of the implicit function theorem first introduced by Kolmogorov for celestial mechanics and by Nash for the embedding problem of Riemannian manifolds. The chapter starts with the simple model problem of differential equations equations on the torus, with the corresponding Poincaré problem for transformations of the circle. In Sections 7 and 8 we state abstract implicit function theorems which imply the Kolmogoroff-Arnold theorem and the Moser twist theorems. In Appendix II we sketch the application of the implicit function theorem to the Nash embedding theorem. As the reader can check for himself, the presentation owes a good deal to the Varenna lectures of Moser [M 4]. However the chapter does not assume any analysis beyond that found in a modern advanced calculus book such as [L 3] . Also included in this chapter are a sketch of the Birkhoff fixed-point theorem and an application of these various theorems to the problem of motion on a convex billiard table (and geodesics on convex surfaces close to spheroids).

Chapter IV presents the classical canonical equations (and reductions) for the three body problem and the restricted three body problem. The exposition follows Poincaré [P 5]

and [P 6] very closely. The subject is brought to the point

where the theorems of Chapter III become applicable. For

the full development of the subject with all its rich side-

lights there is no substitute for reading Poincaré in the

original. In addition, some of the recent work on the

periodic solutions of the restricted three body problem is

presented.

I had initially intended to discuss the celebrated

work of Arnold on the stability of the solar system. How-

ever, due to lack of time and energy we did not get to this

point in the course. Arnold's theorem involves an additional

technique - that of averaging of the fast variables - beyond

the methods discussed in the notes.

In the course, the relation between the various

normal form theorems and conjugacy theorems for Cartan

subgroups was discussed. I have not included this material

in the notes because I feel that it is not yet completely

understood. The reader can refer to the original papers

[S 13], [S 14], and [M 5] . Other subjects that should be

made available are the various structural stability and

mixing theorems of Sinai and Anosov and the general

structure theorems of Smale and his school. A general

survey by Smale [S 7], has appeared. It is to be hoped that
in a short time the subject will have crystallized to the
point where concise proofs of the main results will be
accessible.

When these notes were about two thirds completed
I came into possession of the excellent book by Arnold and
Avez. Frankly, had I seen this book earlier I would, most
likely, not embarked on the task of writing up these notes.
The book by Arnold and Avez is a delightful introduction to
the current ideas in mechanics and very successfully
conveys the flavor of the subject. I strongly recommend
it to one and all.

A comment on prerequisites is in order. Through-
out most of these notes nothing is assumed beyond what
can be found in [L3] . In one or two places (mainly the
appendix to Chapter I) a bit of measure theory is assumed.
At certain points we refer the reader to Hartman's book
[H 2] for further development of various subjects in differential equations. However, the material in [L 3] is very
definitely assumed. In particular, Chapter IV and Section
14 of Chapter III rely somewhat on the treatment of classical
mechanics of Chapter 13 of [L 3] .

My first introduction to the beauties of celestial mechanics came at the hands of Professor Wintner to whose memory these notes are dedicated. In addition to giving me my early mathematical training, he managed to inspire in me a love for the subject which has lasted these many years. I must also mention the inspiring course of lectures that were given by Professor C. L. Siegel at Johns Hopkins during 1955 which I still vividly recall.

Professor Ravindra Kulkarni has provided invaluable assistance in reading the entire manuscript and suggesting many improvements. I must also thank Professor A. B. Novikoff for his reading of portions of the manuscript and suggested improvements. Needless to say, I take sole responsibility for the many faults that still exist. I must also thank Mrs. Mary McQuillin for her excellent typing, editorial and production work.

חושלב ע   ויהי נועם וגו'

Jerusalem

July 1968

CELESTIAL MECHANICS

CHAPTER I.

ALMOST PERIODIC FUNCTIONS AND MEAN MOTION

§ 1. <u>Almost periodic functions</u>. The earliest and
crudest observation of the heavens leads to the conclusion
that celestial motion is cyclic in some approximate sense.
The stars appear to rotate about the heavens once each day.
On closer examination it turns out that the stars lose a
little under four minutes each day so that after about $365 \frac{1}{4}$
days they return to their original position. After some
dispute among the ancients (cf. Pannekoek [P 1]) this was
interpreted as follows: The stars are fixed in a celestial
sphere which is steadily rotating about the earth. The sun
moves in a great circle on this sphere completing its
circuit in about $365 \frac{1}{4}$ days. The period of return of the
sun to its original position is called a year and there is an
observed gross connection between the various positions of
the sun on its course and meteorological conditions, i. e.,

1.

the seasons of the year. A small number of the brighter
stars appear also to be moving on the celestial sphere
(and are therefore called planets) as does the moon . All
of these motions exhibit certain approximate periodicities.
The common situation is, roughly speaking, as follows: to
a crude degree of approximation the planet (or moon) seems
to return to its original position in approximately equal
periods of time. Upon more careful observations it
turns out the planet has not quite returned to its original
position in the approximate periods. However, other
(longer) periods seem to exist in which the planet returns
still closer to its original position.

In no case do the planets or the moon get close to the
poles. Their latitude thus ranges over a bounded open
interval $0 < r < \pi$ while the longitude is, of course, an
angular variable, θ . The position of a planet on the
celestial sphere can thus be described by the complex
number $re^{i\theta}$. Thus the motion of the planet in time can
be described by a complex valued function f . The
approximate regularities and periodicities in the motion
can be idealized by saying that it is an almost periodic
function according to the definition given by Bohr [B 12] :

DEFINITION . A complex-valued function, f, is said to be almost periodic if

i) f is continuous

ii) to every $\epsilon > 0$ there is an $L(\epsilon)$ such that any interval $(a, a + L(\epsilon))$ of length $L(\epsilon)$ contains a τ so that

$$(1.1) \qquad |f(x + \tau) - f(x)| < \epsilon , \qquad \text{for all} \quad x .$$

Condition i) needs no explanation. Condition ii) can be reformulated as follows: Let us call a real number τ ϵ-approximate period of f if (1.1) holds. Condition ii) says that there exist ϵ periods for any ϵ and that the ϵ periods are roughly "of the form nL" where $L = L(\epsilon)$, in the sense that each interval $(nL, (n + 1)L)$ contains an approximate period of f .

The simplest example of an almost periodic function is a periodic one. If $f(x + L) = f(x)$ then nL is an ϵ period for any ϵ . The next simplest example of an almost periodic function is a trigonometric polynomial i. e. , a function of the form

$$f(t) = a_1 e^{2\pi i \lambda_1 t} + \cdots a_k e^{2\pi i \lambda_k t} .$$

Let us prove that such a function is almost periodic. [†] If all the λ's were rational, then f would be a periodic function

[†] The less technically inclined reader can skip the bottom of p. 11.

4.

whose periods are integral multiples of the common denominator of the λ's . Similarly, if all the λ's are rational multiples of μ , i.e.,

$$\lambda_j = \frac{p_j}{q_j} \mu$$

and Q is a common multiple of q_j's then all multiples of Q/μ would be a period of f . In the general case the argument is a little more involved. For any ϵ we would like to find an $L(\epsilon)$ such that every interval of length L contains a τ such that the inequalities

(1.2)
$$|\tau \lambda_j - n_j| < \epsilon \qquad \text{all } j$$

for suitable integers n_j . This is clearly sufficient to establish the almost periodicity of f . As to the necessity if $a_j = 1$ for all i then $f(0) = k$ and any δ almost period would satisfy $|f(\tau) - k| < \delta$. Clearly this can happen only if $|e^{2\pi i \lambda_j \tau} - 1|$ is small for each j and this implies that (1.2) holds.

Now a classical argument of Dirichlet shows that for any ϵ we can find a positive number τ such that (1.2) holds. In fact his argument shows that we can find an integer τ satisfying (1.2) . For the sake of completeness

we reproduce his argument here. Choose the integer K
so large that $K^{-1} < \epsilon$.

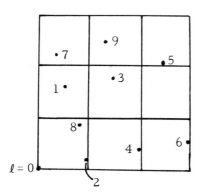

Sub-divide the unit cube into K^k congruent cubes. There
are $K^k + 1$ points of the form $(\{\ell\lambda_1\}, \cdots \{\ell\lambda_k\})$ $0 \le \ell \le K^k$
in the unit cubes, where $\{x\} = x - [x]$ denotes the fractional
part of x (and $[x]$ denotes the largest integer $\le x$) .
Therefore at least two such points lie in the same box, say
$(\{\ell\lambda_1\}, \cdots, \{\ell\lambda_k\})$ and $(\{m\lambda_1\}, \cdots, \{m\lambda_k\})$. Thus

$$|\{\ell\lambda_i\} - \{m\lambda_i\}| < \frac{1}{K} \qquad \text{all i .}$$

Thus $|\ell\lambda_i - m\lambda_i - n_i| < \frac{1}{K}$ for all i where n_i are
suitable integers. We may assume that $\ell > m$ and write
$\ell - m = p$ where $0 < p < K^{k+1}$. We have thus satisfied
(1.2) with $\tau = p$. Observe, however, that this does <u>not</u>
prove that f is almost periodic. In fact, we have produced

only one almost period. We still have to show that there is some $L(\epsilon)$ such that any interval of length L contains an almost period.

We will prove something more general. We shall, in fact, show directly that the sum of two almost periodic functions is almost periodic. For this we shall need some lemmas about almost periodic functions. This proof will be independent of the previous argument. We shall present two proofs, a direct one on pp. 7-9 and a more conceptual one on pp. 10-11.

LEMMA 1.1. An almost periodic function f is uniformly bounded on $\mathbb{R}$.

Proof: Consider $L(1)$. Since f is continuous it is bounded on $[0, L(1)]$, i.e., $|f(x)| \le M$ for $x \in [0, L(1)]$. But then for any t we can find an almost period in the interval $(t - L(1), t)$. Thus $|f(t) - f(t - \tau)| < 1$ so by the choice of M we have

$$|f(t)| \le M + 1 \qquad \text{for all} \quad t .$$

LEMMA 1.2. An almost periodic function is uniformly continuous on $\mathbb{R}$.

Proof: Choose $L(\epsilon/2) > 1$ and then δ such that $|f(x) - f(y)| < \frac{\epsilon}{3}$ for all $|x - y| < \delta$ whenever x, y lie in $(0, 2L(\epsilon/3))$. Thus for arbitrary u, v, we have

$$|f(u) - f(v)| \ \leq \ |f(u) - f(u - \tau)|$$

$$+ \ |f(u - \tau) - f(v - \tau)|$$

$$+ \ |f(v) - f(v - \tau)|$$

$$\leq \ \epsilon \ .$$

Q.E.D.

Let f be an almost periodic function. We set
$L(f, \epsilon)$ = length of the interval containing an ϵ-period and
$\mathcal{S}(f, \epsilon) = \{\tau \in \mathbb{R} \,|\, \tau$ is an ϵ-almost period of $f\}$.

LEMMA 1.3.   Let f, g be almost periodic functions.
Then   1) $\{\tau_1 + \tau_2 \,|\, \tau_i \in \mathcal{S}(f, \epsilon_i) \ \ i = 1, 2\} \subseteq \mathcal{S}(f, \epsilon_1 + \epsilon_2)$

2) $\mathcal{S}(f, \epsilon_1) \cap \mathcal{S}(g, \epsilon_2) \subseteq \mathcal{S}(f + g, \ \epsilon_1 + \epsilon_2)$.

3) Given $\epsilon > 0$, there exists $\delta > 0$ such that

$$\{\tau \,|\, |\tau| \leq \delta\} \subseteq \mathcal{S}(f, \epsilon) \ .$$

4) Given $\epsilon_1, \epsilon_2 > 0$, there exists $\delta > 0$ such that

$$\{\tau + \eta \,|\, \tau \in \mathcal{S}(f, \epsilon_1), \ |\eta| \leq \delta\} \subseteq \mathcal{S}(f, \ \epsilon_1 + \epsilon_2) \ .$$

Proof: 1) and 2) are easy. 3) follows by uniform
continuity of f established in Lemma 1.2. To prove 4),
use 3) to choose δ such that every η , with $|\eta| \leq \delta$, is

an ϵ_2-period of f . Then 4) follows by 1) .   Q.E.D.

Now we shall show that the sum of two almost periodic functions is almost periodic:

Indeed, let f and g be almost periodic, and fix $\epsilon > 0$. Let $L \geq \{L(f, \epsilon/6) , L(g, \epsilon/6)\}$, and $\delta > 0$ so small that every η , with $|\eta| \leq \delta$, is an $\epsilon/3$-period of both f and g .

Divide the interval $[0, L]$ into subintervals I_i of length $\leq \delta$. Some sub-interval I_i may contain differences $|\tau_f - \tau_g|$. For these, pick a fixed pair τ_f^i , τ_g^i such that $|\tau_f^i - \tau_g^i| \in I_i$. Let $T \geq$ all τ_f^i , τ_g^i .

We claim that every interval of length $L + 2T$ contains an ϵ-period of $f + g$. Indeed given an interval of length $L + 2T$, fix the "middle"

interval of length L . By the choice of L , there exists a τ_f (resp. τ_g) belonging to this interval which is an $\epsilon/6$-almost period of f (resp. g) . Then $|\tau_f - \tau_g| \leq L$. Since $|\tau_f - \tau_g| \in I_i$ for some I_i of the sort considered above, then $\tau_f - \tau_g = \pm(\tau_f^i - \tau_g^i) + \eta$ with $|\eta| \leq \delta$, or

$$\tau_f \pm \tau_f^i = \tau_g \pm \tau_g^i + \eta \ .$$

This number obviously belongs to the given interval of

length $L + 2T$. On the other hand, using Lemma 1.3, we

see that $\tau_f \pm \tau_f^i$ is an $\epsilon/3$-period of f , and $\tau_g \pm \tau_g^i + \eta$

is an $2\epsilon/3$-period of g ; consequently, it is an ϵ-period

of $f + g$.

This shows that $f + g$ is almost periodic, and

finally proves, by trivial induction, that

$$f(t) = a_1 e^{2\pi i \lambda_1 t} + \cdots + a_k e^{2\pi i \lambda_k t}$$

is almost periodic.

So far our treatment has been quite "lowbrow".

From a "structural" point of view, however, it will be

useful to reformulate the definition of almost periodicity

in a different way. This reformulation will yield the above

result very simply; it can also be used to define almost

periodicity of a complex-valued function on an arbitrary

group ; and it is also useful in deriving the

fundamental (Bohr's) theorem on almost periodic functions

(cf. the appendix) .

THEOREM 1.1. <u>A continuous function</u> f <u>is almost periodic if and only if the family of functions</u> $\{f(\cdot + h)\}$ <u>is sequentially pre-compact.</u>

<u>Proof</u>: To say that $\{f(\cdot + h)\}$ is sequentially pre-compact means that given any sequence h_i we can choose a subsequence h_{i_j} such that $f(\cdot + h_{i_j})$ is a Cauchy sequence in the uniform norm, i.e., that $f(t + h_{i_j})$ converges uniformly in t . Suppose that f is almost periodic. An almost period of f is clearly an almost period for $f(\cdot + h)$. The proofs of Lemmas 1.1 and 1.2 show that the family $\{f(\cdot + h)\}$ is uniformly bounded and uniformly equicontinuous. It is therefore sequentially precompact by Arzela's theorem. (The proof proceeds as follows: Choose a countable dense set of points, D, on $\mathbb{R}$, say the rationals. By a diagonal procedure, choose a subsequence such that $f(x + h_{i_j})$ is Cauchy at each x and D. By the equicontinuity the sequence $f(\cdot + h_{i_j})$ is Cauchy.)

Conversely, assume f is <u>not</u> almost periodic . Then there is some ϵ and a sequence of intervals I_j whose lengths $|I_j|$ tend to infinity such that I_j contains no ϵ-almost period of f. We must produce a sequence h_i such that $f(\cdot + h_i)$ has no convergent subsequence. Let $h_1 = 0$. Choose h_2 so that $h_2 = h_2 - h_1 \in I_1$. Now choose

k_1 so that $|I_{k_1}| > |h_2 - h_1|$. Then we can choose h_3 such that $h_3 - h_2 \in I_{k_1}$. Proceeding in this way we can arrange that $h_i - h_j \in I_{k_i}$ for all $j \leq i$. Thus for any i and j we conclude that $h_i - h_j$ is not an ϵ almost period of f . Thus

$$\sup_x |f(x + h_i) - f(x + h_j)|$$
$$= \sup_t |f(t + h_i - h_j) - f(t)| > \epsilon$$

and thus no subsequence of $\{f(\cdot + h_i)\}$ is Cauchy.

COROLLARY: Let F be a uniformly continuous function of k real variables. If $f_1, \cdots, f_k$ are almost periodic functions then $F(f_1(\cdot), \cdots, f_k(\cdot))$ is an almost periodic function. In particular, $f_1 + f_2$ and $f_1 \cdot f_2$ are almost periodic if f_1 and f_2 are.

In fact, if each of the $\{f_i(\cdot + h)\}$ are sequentially precompact so is $\{F(f_1(\cdot + h), \cdots, f_k(\cdot + h))\}$.

This last corollary suggests an interesting class of functions: the quasi-periodic functions:

DEFINITION 1.2. A function f(t) is called quasi-periodic if it is of the form

$$f(t) = F(e^{2\pi i \lambda_1 t}, \cdots, e^{2\pi i \lambda_k t})$$

for some continuous function F and real numbers $\lambda_1, \cdots \lambda_k$.

Another way of formulating the definition is to say that f is quasi-periodic if

$$f(t) = G(\lambda_1 t, \cdots, \lambda_k t)$$

where $\rightsquigarrow (\lambda_1 t, \cdots, \lambda_k t)$ is a one parameter group (a straight line) on the k-dimensional torus and G is a continuous function on the torus. (The relation between the two definitions is clearly given by $G(\theta_1, \cdots, \theta_k)$ = $F(e^{2\pi i \theta_1}, \cdots, e^{2\pi i \theta_k})$ where the θ_i (mod 1) are linear coordinates on the torus.

(In the older literature a quasi-periodic function was called conditionally periodic.)

Every quasi-periodic function is clearly almost periodic but not conversely. In the appendix to this chapter we shall show that

PROPOSITION 1.1. Every almost periodic function can be approximated uniformly by quasi-periodic functions.

Since any continuous function on the torus can be approximated uniformly by trigonometric polynomials we conclude that

PROPOSITION 1.2. (Bohr's theorem). <u>Every almost periodic function can be approximated uniformly by functions of the form</u>

$$a_1 e^{2\pi i \lambda_1 t} + \cdots + a_n e^{2\pi i \lambda_n t} .$$

Now a function of this form can be considered geometrically in the plane as a particle rotating uniformly with angular velocity λ_n on a circle whose center is rotating with angular velocity λ_{n-1} on a circle etc. Here $|a_i|$ represents the radius of the i-th circle and $a_i/|a_i|$ represents the initial angular position on the i-th circle at time $t = 0$.

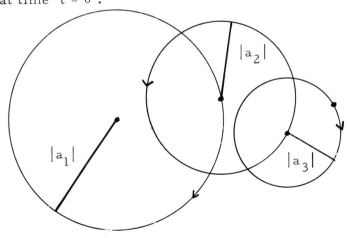

The inner circle is known as the deferent and the outer circles are known as epicycles and Bohr's theorem shows that the motions of the planets (if they are almost periodic) can be represented to any desired degree of approximation by a superposition of epicycles. This was the basic method of Hipparchus and Ptolemy.

§2. The mean month and year. In actual fact the ancient astronomers made some modifications in this method. We shall illustrate their techniques in dealing with the motion of the sun and the moon. In ancient times the only possible way of establishing a calendar (necessary at the very least for legal transactions and religious observances) was to use the natural periodicities of readily observed phenomena, cf. Gen. 1.14. Of these, the most obvious unit was the day and the second most obvious unit was the month - i.e., a periodicity of the moon's motion. It was widely accepted that a new month begins at the instant of visibility of the new lunar crescent, as this was a phenomenon which could be widely observed and would (at least in those regions with no clouds) normally lead to a confusion of at most one day.

The third unit was the year, the return of the sun and the seasons. The moon's period is not an integral multiple of a day, nor is the sun's period an integral multiple of either the day or the month. Since it would be unwieldy to have three independent units in the calendar the month was taken as an integral number of days to be described by the courts.   Furthermore, that the year should be a multiple of months, 12 or 13, to be decided each year by the court. In order for the courts to deal with the calendar in an effective way, they first of all needed to know the average length of the month. This was necessary, for example, during the rainy seasons when the moon might not be visible for several months on end due to meteorological conditions. Under these circumstances the month would have to be established by court decree alone. Cf., for example Maimonides Code

הלכות קדוש החדש פרק י"ח ה'ה-ט (Laws of the santification of the new moon Ch XVIII 5-9.)

      The average length of a synodic period (the interval from one conjunction to the next) was something that could be very readily measured, provided that records be kept for a long period of time. In fact, when an eclipse of the sun occurs, it occurs at conjunction. (Not every conjunction yields an eclipse because the sun, moon and earth don't move in a fixed plane.) If records of eclipses are kept over a period of centuries, then the mean synodic month is obtained by dividing the length of time elapsed between two distant eclipses by the number of intervening months. The result, found in the old Jewish tradition and found on Babylonian tablets dating back several centuries before the common era give the figure 29 days $12\frac{793}{1080}$ hours. (A likely reason for dividing the hour up into 1080 parts is that 1080 has many useful divisors (and so simplifies the arithmetic) and in terms of which the mean synodic month is very closely an integer. For instance, in terms of our unit of one second $= \frac{1}{3600}$ hr. the mean motion would be 29 days 12 hrs. 44 min. $3\frac{1}{3}$ sec. and so is not an integer: cf. p. 37   ספר העבור. For an alternative explanation of the origin of this unit see Professor Neugebauer's astronomical commentary to the English translation of Maimonides code.

The length of the period of the sun was not so easy to determine with great accuracy. (It was not so essential to know the exact length of the year so long as the number of months in a year was determined by the courts each year. The introduction of a fixed calendar, however, required an exact knowledge of the length of the year.)

There were several methods available to the ancients to measure the length of the year. One procedure was to use a vertical stick and measure the length of the shadow. They could then determine the summer solstice (the moment of minimum shadow length) and the winter solstice (the maximum) and the equinoxes (the mid-points). This measurement is very difficult to perform accurately since the variation of the shadow length is very slow near the end points.

Another measurement that was made was the length of the period of return of the sun to the same position among the fixed stars. This is also difficult to measure accurately. (The most accurate measurement could be obtained during a lunar eclipse when the sun's position is opposed the moon's.)

In point of fact these two procedures measured two different things. The first measured the tropical year, i. e., the return of the seasons or the return of the sun to the equinox. The second measured the sidereal year - the return of the sun to the same position relative to the stars. Actually, the position of the sun at equinox relative to the stars is not quite the same but varies from year to year. This was known as the precession of the equinox.

18.

We are unsure at present whether this phenomenon was known to the Babylonian astronomers. It was one of the striking discoveries of Hipparchus.

§3. The motion of the sun. One of the results that was obvious from the shadow length measurements was that the seasons are not of constant length. Thus spring is considerably longer than fall. What this demonstrates is that the sun does not move with constant angular velocity on the ecliptic. This could also be observed directly when charting the sun's motion relative to the fixed stars.

In the earliest Babylonian tables the motion is described as if the sun has two constant velocities during the year. That is, the tables listing the positions of the sun show a constant change, about $30°$ (mean) per month during one portion of the ecliptic (from longtitude $163°$ to $357°$) and a change of $28 \frac{1}{8}°$ per month on another (from $357°$ to $163°$). In a sense this must be regarded as one of the earliest formulations of a physical law and prediction on record. Most likely, the two velocity hypothesis was assumed. Then the length and position of each constant velocity period, together with the value of the velocity was constructed from the shadow length measurements . From these the tables were then constructed. Thus

(most likely) the hypothesis concerning the nature of the
sun's motion was used to predict its position in the tables.

Hipparchus used the method of the eccentric circle
to explain the variation in the sun's velocity. The basic
assumption is that the sun moves on the ecliptic at a
constant angular velocity about the center of the ecliptic,
but the center of the ecliptic does not coincide with the
earth.

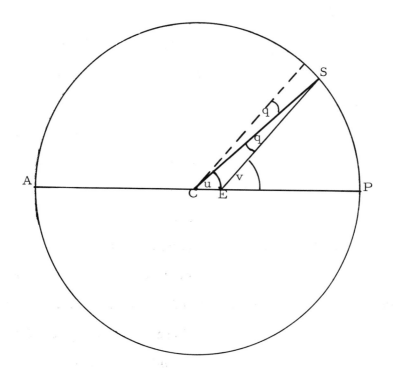

C   is the center of rotation
E   is the earth, A is apogee ,   P is perigee .

The sun S moves with a uniform angular velocity about the circle. This describes the mean motion of the sun. The apogee, A, also describes a uniform but very slow rotation (about 1° in 70 years) due, in part, to the precession of the equinox. The motion of the apogee (beyond the effect due to precession) was apparently first discovered by the Arabian astronomer Al-Battani. The angle u from the sun to perigee through C is called the (eccentric) anomaly.

To illustrate how the position of the sun was calculated in medieval texts we outline the procedure followed in Maimonides' code. First the position of the mean sun is recorded at some starting date, (Chapter XII Halacha 4). The rate of the sun's mean motion is given (Chapter XII h. 1), as is the rate of the motion of the apogee (Chapter XII, h. 3). Thus the angle u may be computed for any desired date (Chapter XIII, h. 1). Now the angle $q = v - u$ is called "the quota of the anomaly". Its value is tabulated as a function of u for intervals of 10° from 0° to 180° (Chapter XIII, h. 4). For $180° \le u \le 360°$ we clearly have $q(u) = -q(360° - u)$(Chapter XIII h. 2, 5 Maimonides measures angles from P rather than A so there is a shift of 180°.) For values of u which are not multiples of 10°, q(u) is computed by linear interpolation

(Chapter **XIII**, h. 7). The value q(u) is added to the mean position of the sun to obtain the true position of the sun.

Notice that Hipparchus' theory not only gives an empirical law for describing the motion of the sun, it also makes an assertion relating the velocity of the sun to its distance from the earth: the angular velocity, dv/dt , is greater when the sun is closer to earth and smaller when the sun is far away. This in turn should have the effect of changing the apparent size of the sun's disk, i. e., the sun should look larger at **A** and smaller at P . This phenomenon is indeed observed and was mentioned by the greek astronomers as verification of this explanation of the variability of solar velocity.

Let us compute this effect. Take the x-axis along **PA** and the center of coordinates at **E** . Let the radius **CS** be a and CE = ea . If the sun's coordinates are x, y then

$$y = a \sin u$$
$$x = a(\cos u - e)$$

while

$$\tan v = \frac{y}{x} .$$

22.

Thus

$$\frac{1}{\cos^2 v} \frac{dv}{dt} = \frac{1 - e \cos u}{[\cos u - e]^2} \frac{du}{dt}$$

and setting $\frac{du}{dt} = k$ we see that the maximum velocity is

achieved when $u = v = 0$ and is

$$v_{max} = \frac{K}{1 - e}$$

while

$$v_{min} = \frac{K}{1 + e}$$

is achieved at $v = u = \pi$. Thus

$$\frac{v_{max}}{v_{min}} = \frac{1 + e}{1 - e} = 1 + 2e + \cdots$$

for small values of e . Letting d be the apparent solar

diameter we clearly have $2 \tan \frac{1}{2} d = \frac{S}{\text{distance ES}}$ where

S is the actual

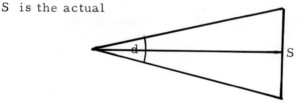

diameter of the sun. Thus

$$2 \tan \frac{1}{2} d_{max} = \frac{S}{a(1 - e)}$$

$$2 \tan \frac{1}{2} d_{min} = \frac{S}{a(1 + e)} \quad .$$

For small values of d we have, approximately $\tan d \cong d$

so that

$$\frac{d_{max}}{d_{min}} \sim \frac{1 + e}{1 - e} \quad .$$

Thus in the Hipparchus theory we should have d_{max}/d_{min}

approximately equal to v_{max}/v_{min} or, since e is small,

$$\frac{v_{max}}{v_{min}} - 1 \sim 2e \sim \frac{d_{max}}{d_{min}} - 1 \quad .$$

Now in the more exact Kepler theory of elliptical

motion we have

$$\frac{k^v max}{k^v min} = \left(\frac{1 + e}{1 - e}\right)^{1/2} \sim 1 + e \quad .$$

Thus there is a discrepency of a factor of two in

$\dfrac{d_{max}}{d_{min}} - 1$ in the theory of Hipparchus. Unfortunately,

the apparent solar diameter was something very difficult

to measure. Therefore the qualitative effect was taken as

a verification of the theory while the quantitative refutation

was not possible with their experimental equipment. As

we have already indicated, an exact value for the tropical

or sidereal year was not essential to the ancient courts so

long as the length of the year was determined by the court from year to year. In fact, the value of the tropical or sidereal year was consistently overestimated by from 4 to 10 minutes. (The rather exact looking value given by Maimonodes in Chapter X , h. 1 , the תקופה דרב אדא בר אהבה (period of R. Adda bar Ahaba) is an artificial figure based on the fixed 19 year = 235 month cycle calendar rather than astronomical fact. Thus the year (for the purpose of this calendar) was defined as $\frac{1}{19}$ of the cycle of mean months. This explains the strange subdivision of $\frac{1}{1080}$ hr. into $\frac{1}{76}$ submits. The fraction $\frac{1}{76}$ was obtained (for computing the mean season) as $\frac{1}{4} \cdot \frac{1}{19}$. This value of the year does not coincide with the astronomical value corresponding to the mean motion of the sun and precession of the equinoxes given by Maimonides in Chapter XII .)

§4. The motion of the moon. The fundamental astronomical problem that concerned the courts was to determine the date of first visibility of the lunar crescent after new moon. Furthermore they were concerned with the shape of the lunar crescent at first visibility so that they could cross-examine witnesses who testified to the observation of the new moon.

The two main factors that enter into the problem of visibility are the angular distance from the sun to the moon (which determines the size of the lunar crescent) and the height of the moon above the horizon (which determines how dark it will get before the moon sets). This latter effect is governed not only by the angular distance from the sun to the moon (along the ecliptic) but also on whether the moon is north or south of the ecliptic and the angle the ecliptic makes with the horizon. That the angle the ecliptic makes with the horizon can have a striking effect at full moon is shown by the phenomenon of the harvest moon. At the fall equinox, the sun is crossing the equator heading south. The full moon, at the opposite side of the ecliptic, is headed northward. This northerly motion of the moon has the effect of decreasing the delay of moonrise. At spring the opposite effect is true. Thus at the latitude of New York, the delay (at full moon) of moonrise in September may be as little as 15 minutes while it may be as great as 1 h^2 15 minutes in March.

These matters, together with the effect of parallex are dealt with in Chapter XVII and XVIII of Maimonodes code. As they are not central to our story we shall not discuss them.

In any event, the crucial effects to be determined
are the moons motion parallel to the ecliptic (the longitude)
and its deviation above and below the ecliptic. The moon's
motion relative to the stars can be readily plotted. The
return to the same position relative to the fixed stars is
called the sidereal month. It is readily observed that the
lunar velocity is very far from being constant. The moon's
movement of longitude varies from about 11° to about 15°
per day. It is discovered that the moon has approximately
one maximum velocity per month. In fact, the return to
maximum velocity takes slightly longer than a sidereal
month and slightly less than a synodic month. The period
of return to maximum velocity is known as the anomalistic
period or anomalistic month. Since the mean (synodic)
month is known, the anomalistic month can be obtained by
counting the number of maximum velocities in a large
number of (synodic) months. Hipparchus gives the relation
that 251 synodic periods are almost exactly equal to 269
anomalistic periods.

What is most striking is that we find this ratio ,
251/269 used in a theoretical way in Babylonian tables.
In fact, the effect of the moon's variable velocity is to

change the length of the actual (as opposed to mean)

synodic month. We find Babylonian tables giving the

length of successive months. It is noticed that the lengths

steadily increases $1^h 30^m$ per month for awhile and

steadily decrease at a rate of $1^h 30^m$ per month. The

only place where this steady increase or decrease of

$1^h 30^m$ is not observed is at the turning points. The

ingeneous and obviously correct solution to the puzzle of

how the tables were constructed was solved by the scholars.

It turns out that the figures are exactly the values of a

function whose graph is given below.

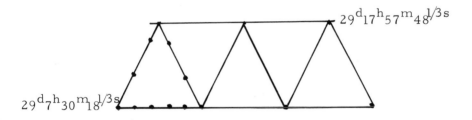

$29^d 17^h 57^m 48 \frac{1}{3} s$

$29^d 7^h 30^m 18 \frac{1}{3} s$

That is, a function whose slope alternates between $+1^h 30^m$

and $-1^h 30^m$ and whose extremes are those shown on the

graph. Each unit is a month and one discovers that there

are a little less than 14 units in each period of the function.

More precisely, it turns out that 18 periods contain 251

months. Assume that the relative position of the new

moon with regard to length of the month will coincide with

the return of period of anomalistic month to the synodic month, and that the synodic month is slightly longer than the anomalistic. Then this says that the moon has passed an extra 18 periods after 251 months. Thus we obtain 269 anomalistic periods for every 251 synodic months. We see clearly that the Babylonians postulated a specific type of function to describe the moon's motion. This type of function is common in the late Babylonian tables of the motion of the various planets. The choice of slope and end points were most likely determined by the anomalistic period.

The variation in the moon's latitude (relative to the ecliptic) was also irregular. It was between +5° and -5°. In this case, the Babylonians were aware that the curve describing the latitude could not be a simple zigzag curve as on the previous page. Such a curve would have too small a slope near lat = 0 and at this point the slope is crucial since an eclipse will occur if the latitude is sufficiently small at new or full moon. Apparantly they used a polygonal curve which had the shape given below.

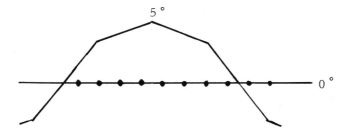

Hipparchus used an eccenter to describe the motion of the moon while Ptolemy used the method epicycles. We describe a simplified version of Ptolemy's theory following O. Neugebauer [N 2] .

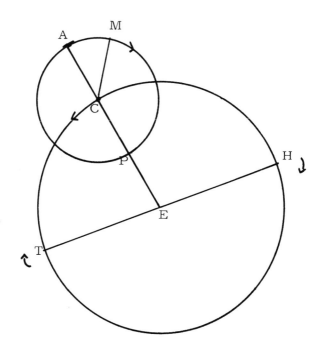

The plane of the moon's orbit is inclined to the ecliptic at an angle of about 5°. The line of intersection of these planes is called the line of nodes. This line rotates slowly in a retrograde movement (i.e., an east to west direction). The moon rotates about the small circle with center C and this center rotates on the large circle. The point H denotes the ascending node (the dragon's head)[†] where a rotating C moves north of the ecliptic while T is the descending node (the dragon's tail). The period of rotation of M about the small circle is slightly less than that of C around the large circle so that the return to P (the maximum velocity) takes a little longer than a return to C. The rotation of M about C is called the mean (motion of the) anomaly. This picture approximately the motion of the moon both in latitude and in longitude. It turns out that while this scheme is fairly accurate near conjunction and opposition (new moon and full moon) it deviates from the observed positions of the moon at quadrature. In fact, the maximum observed deviation from mean motion is about 5° at new moon and full moon and about 7°40' at quadrature. Thus two corrections were introduced depending on the (mean) moon's position relative to the (mean) sun.

[†] This notation is a carryover from a previous theory in which the moon was assumed to be swallowed by a dragon in an eclipse.

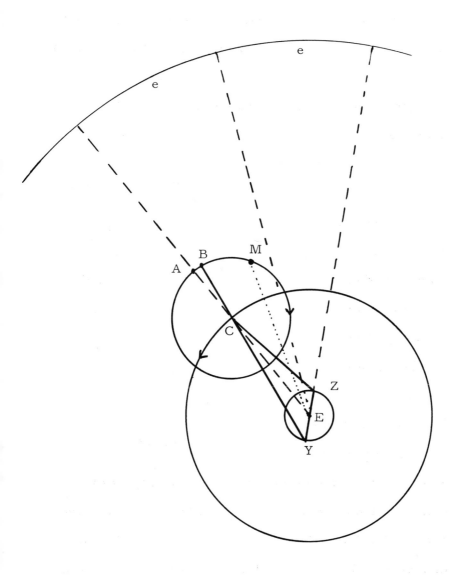

It is assumed that the center of the epicycle, instead of being equidistant from the earth, is equidistant from a point Z which has retrograde motion on a small circle about the earth. The point Z is such that the angle CEZ = 2 angle CES. The angle CES is known as the (mean) elongation and so CEZ is called the double elongation. This means that the curve described by C around E is really an oval whose long axis points towards the sun.

This has the effect of increasing the apparent angular diameter of the epicycle at quadrature. This explained the difference in maximum observed deviation. This effect is known as the evection. Another

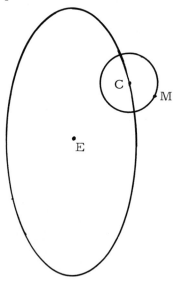

small correction was that the mean anomaly was to be computed from the point B (which is obtained by inter-secting the line YC with the epicycle where YZ is a diameter of the small circle). The true anomaly is then obtained by adding AB to BM .

Thus the procedure for finding the position of the moon is: (1) Compute the mean moon, the mean sun and the mean anomaly. (2) Compute the double elongation and determine AB as a function of it. Actually AB is tabulated as a function of the double elongation (Chapter XV h3 in Maimonides). (3) Adding AB to the mean anomaly to obtain the true anomaly we can then compute the moon's position as a function of the true anomaly and elongation. These functions are tabulated. (Actually Maimonides assumes a fixed distance from E. That is, he tabulates for an average elongation as he is only interested in the moon's position near a position of possible first visibility). (4) Locate the position of the dragon's head and tail and find the relative position of the true moon to the head or tail. (5) Find the latitude of the moon. This latitude is tabulated as a function of the angular distance from the true moon to H (or T).

This was Ptolemy's explanation of the motion of the moon. Tycho Brahe discovered that while Ptolemy's theory accounted very closely for the facts near conjunctions oppositions and quadratures, it could deviate by as much as 1° at octants (45° or 135° to the sun). He therefore introduced another correction known as the variation.

Despite the fact that the theory outlined above gives a fairly accurate description of the moon's motion on the celestial sphere, it could not be accepted as describing actual spatial relations between the earth and the moon. This is in contrast to what we described earlier about the motion of the sun. In fact, since the deviation from mean motion is considerable, both the radius of the epicycle and of the circle carrying Z are non-negligible. As a consequence, it turns out that the maximum distance of the moon from earth (at apogee at conjunction or opposition) is almost twice as large as the minimum distance (at perigee at quadrature). This means that some quarters, the moon's diameter should appear twice as large as at some full moons and this is never observed.

§5. The planets. It is the theory of the motion of the planets that represents the greatest achievement of Ptolemy. Here, by a clever superposition of many epicycle motions in space he was able to approximate very closely the motions of the planets. Complete tables are given for many of the interesting astronomical phenomenon as deduced from the theory. We will not describe this theory in detail but content ourselves with a few general remarks. For simplicity we will discuss the situation with one

epicycle although our remarks hold true in general. Let
w(t) represent the motion of the center of the epicycle
(where $w(t) \in \mathbb{R}^3$) . Let v(t) represent the motion on the
epicycle. Then

$$x(t) = v(t) + w(t)$$

is the motion of the planet. Since the right hand side is
symmetric in v and w we see that the roles of epicycle
and deferent can be interchanged.

Secondly, all that is observed is the projection
onto the celestial sphere. Thus all radii can be multiplied
by the same factor without changing the result. In particular, while the ratio of the radii of the epicycle to deferent
can be computed, there is no way of determining any
relationship among radii of different planets.

In Ptolemy the deferent of a superior planet is
traversed in a year while the epicycle of an inferior planet
is traversed in a year. Interchange the circles so that all
deferants are traversed in a year. Normalize so that the
deferents of all planets coincide. Thus all the M's move
on the same circle. It is discovered that all the M's
coincide and coincide with the sun. Thus all the planets
revolve around the sun and the sun revolves about the
earth. This is the system of Tycho Brahe.

APPENDIX : CHAPTER I

MEAN MOTION, THE ERGODIC THEOREM
AND THE BOHR APPROXIMATION THEOREM

§1.   The problem of mean motion.   As we have
already indicated, a trigonometric polynomial

(1.1) $z(t) = a_1 e(\lambda_1 t) + \cdots + a_n e(\lambda_n t)$

where $e(\theta) = e^{2\pi i \theta})$ arises in planetary theories of Ptolemy.
It also arises in the theory of small vibrations and in the
theory of secular variations of Lagrange.   Lagrange raised
the question of whether or not $z(t)$ possesses a mean
angular motion.   That is, let us write

(1.2) $z(t) = r(t) e^{i\varphi(t)}$

where φ is a continuous function.   (Assume for the moment
that $z(t)$ is nowhere zero .)   Can we find an asymptotic
expression for $\varphi(t)$?   More precisely, is

36.

(1.3) $$\varphi(t) = \mu t + o(t)$$

for some constant μ ?

Notice that if

$$|a_1| > |a_2| + \cdots + |a_n|$$

the angle at the origin from $z(t)$ to $a_1 e(\lambda_1 t)$ is never more than $\pi/2$. Thus in this case we can write

$$\varphi(t) = \lambda_1 t + O(1) .$$

In other words λ_1 is the mean motion of the "planet" z . This corresponds to the situation in the Ptolemaic system.

Lagrange himself observed that the case of two epicycles with equal radii also have a mean motion. Here some care must be taken with the definition since z becomes zero. Lagrange noticed that

$$ae(\lambda_1 t + \alpha_1) + ae(\lambda_2 t + \alpha_2)$$

$$= a\left\{ e\left(\frac{1}{2}[\lambda_1 - \lambda_2]t + \frac{1}{2}[\alpha_1 - \alpha_2]\right) + e\left(\frac{1}{2}[\lambda_2 - \lambda_1]t + [\alpha_2 - \alpha_1]\right) \right\} \cdot$$

$$\cdot e\left(\frac{1}{2}[\lambda_1 + \lambda_2]t + \frac{1}{2}[\alpha_1 + \alpha_2]\right)$$

$$= 2a \cos \pi([\lambda_1 - \lambda_2]t + [\alpha_1 - \alpha_2]) \, e\left(\frac{1}{2}[\lambda_1 + \lambda_2]t + \frac{1}{2}[\alpha_1 + \alpha_2]\right) .$$

Thus setting $\varphi(t) = \frac{1}{2}[\lambda_1 + \lambda_2]t + \frac{1}{2}[\alpha_1 + \alpha_2]$ we see that

$$z(t) = r(t)e(\varphi(t))$$

where r now may take on negative values. Here φ possesses a mean motion which is $n = \frac{1}{2}[\lambda_1 + \lambda_2]$. Thus in the limiting case $a_1 = a_2$ the mean motion is the average $\frac{1}{2}[\lambda_1 + \lambda_2]$.

To explain the reason for allowing negative r, let $z(t)$ be any function of t and suppose that $z(t_0) = 0$ while $z^{(k)}(t_0) \neq 0$ for some k. Let ℓ be the first non-vanishing derivative of z at t_0. Then the mean value theorem asserts that

$$z(t) = (t - t_0)^\ell z(\tau) \qquad \text{where } \tau \text{ lies between t and } t_0.$$

Thus

$$\frac{z(t)}{|z(t)|} = \frac{(t - t_0)^\ell}{|t - t_0|^\ell} \frac{z^{(\ell)}(\tau)}{|z^{(\ell)}(\tau)|} .$$

This shows that we can write

$$z(t) = r(t)e(\varphi(t))$$

where φ is a continuous function provided that we make $r(t)$ change sign if ℓ is odd. In this way we obtain a continuous (and differentiable) function φ which is now determined up to π .

Notice that if $z(t)$ is a trigonometric polynomial it is real analytic and thus all derivatives cannot vanish at t_0 unless $z \equiv 0$. (In fact, it is easy to see that not more than k derivatives vanish if these are exactly k distinct λ's mod 1.)

The problem posed by Lagrange is to determine whether a mean motion exists for a general polynomial . The answer to this question is in the affirmative as was first proved by Hartman [H3]. A discussion can be found in a long paper by Jessen and Tornehave [J1], where many other interesting problems are dealt with. Here we shall present the very elegant argument of H. Weyl [W6] which works in the "generic" case and gives a formula for the mean motion, which was first conjectured by Wintner .

§2. Wintner's formula. Writing $z = x + iy$ and

$$z(\theta) = a_1 e(\theta_1) + \cdots + a_n e(\theta_n)$$

we see that the Jacobian matrix of this map of the torus into the z plane is

$$
\begin{pmatrix} \dfrac{\partial x}{\partial \theta_1} & \cdots & \dfrac{\partial x}{\partial \theta_n} \\[2mm] \dfrac{\partial y}{\partial \theta_1} & \cdots & \dfrac{\partial y}{\partial \theta_n} \end{pmatrix} = \begin{pmatrix} -a_1 \sin \theta_1 & \cdots & -a_n \sin \theta_n \\[2mm] a_1 \cos \theta_1 & \cdots & a_n \cos \theta_n \end{pmatrix}.
$$

Assuming (as we may) that none of the a_k vanish, we see that the Jacobian matrix has rank < 2 only at points when $\theta_i - \theta_j = k_{ij}\pi$ for integers k_{ij}. If this is to happen at some point where $z = 0$ we must have $\Sigma \pm a_k = 0$. Thus if $(a_1, \cdots, a_n)$ avoids a finite number of hyperplanes we conclude that $z^{-1}(0)$ is a manifold of dimension $n - 2$. In what follows we shall assume the $z^{-1}(0)$ is in fact of codimension (at least) two. (Thus we are excluding the case considered by Lagrange.)

We will also assume the λ's are linearly independent over the rationals. According to the

KRONEKER-WEYL THEOREM. _If_ f _is any_ Riemann integrable function on the torus,

$$
\lim_{T \to \infty} \frac{1}{T} \int_0^T f(t \circ \lambda + \alpha)dt = \int f(\theta)d\theta
$$

where $\lambda = (\lambda_1, \cdots, \lambda_n)$, $\alpha = (\alpha_1, \cdots, \alpha_n)$, $\theta = (\theta_1, \cdots, \theta_n)$ _and_ $d\theta = d\theta_1 \cdots d\theta_n$. (_If we take_ f _to be the characteristic_ function of a set E then this theorem asserts that the

average length of time the particle spends in E equals the

measure of E.)   We will present a proof of this theorem

below.

For the sake of simplicity let us assume, temporarily,

that the constants α are chosen so that the line $t\lambda + \alpha$

does not intersect the set $z^{-1}(0)$.   This is a generic choice

since dim $z^{-1}(0) \leq m - 2$.   Then

$$z(\lambda t + \alpha) \ = \ r(t)e(\varphi(t))$$

defines a differentiable function φ for all time . Any two

functions φ differ by a constant multiple of 2π .   Further-

more

$$\varphi'(t) \ = \ \mathrm{Re} \ \frac{1}{2\pi i} \ \frac{\frac{dz}{dt}(\lambda t + \alpha)}{z(\lambda t + \alpha)}$$

does not depend on the choice of φ .   Now

$$\frac{\varphi(t_2) - \varphi(t_1)}{t_2 - t_1}$$

$$= \ \frac{1}{t_2 - t_1} \int_{t_1}^{t_2} \mathrm{Re} \ \frac{\lambda_1 a_1 e(\lambda_1 t + a_1) + \cdots + \lambda_n a_n e(\lambda_n t + \alpha_n)}{a_1 e(\lambda_1 t + a_1) + \cdots + a_n e(\lambda_n t + \alpha_n)} dt \ .$$

If we could apply the Kronecker-Weyl theorem to the

function

$$\Phi(\theta) \;=\; \mathrm{Re}\; \frac{\sum_j \lambda_j\, a_j(\theta_j)}{\sum_j a_j(\theta_j)}$$

we would conclude that the mean motion exists and is given by

(1.4) $$\int \Phi \;=\; W_1 \lambda_1 + \cdots + W_n \lambda_n$$

where

(1.5) $$W_i \;=\; \int \mathrm{Re}\; \frac{a_i\, e(\theta_i)}{\sum_j a_j\, e(\theta_j)}\; d\theta \quad .$$

Since the function Φ becomes infinite on $z^{-1}(0)$ this theorem doesn't apply directly. Nevertheless, let us assume that the conclusion holds and compute the integral in question.

Let us compute W_n by first integrating with respect to θ_n holding the remaining variables fixed. Let $b = \sum^{n-1} a_k e(\theta_k)$ and

$$w(\theta_n) \;=\; b + a_n e(\theta_n) \quad .$$

Then the integrand becomes

$$\mathrm{Re}\left(\frac{1}{2\pi i}\; \frac{w'(\theta_n)}{w(\theta_n)}\right) \quad .$$

Integrating this yields zero or one according to whether the circle described by $w(\cdot)$ does or does not contain the origin. In fact, writing $w(\theta_n) = r(\theta_n)e^{i\varphi(\theta_n)}$ this integrand is exactly $\varphi'(\theta_n)$.

Now the circle described by w will contain the origin if $a_n > |b|$ and will not if $|b| > a_n$. Thus

$$(1.6) \qquad\qquad W_i = \mu(E_i)$$

where E_i denotes the set of those values of $\theta_1, \cdots, \theta_{i-1}$, $\theta_{i+1}, \cdots \theta_n$ for which

$$\left| \sum_{j \neq i} a_j \, e(\theta_j) \right| < a_i$$

and $\mu(E)$ is the measure of E .

The geometric meaning of the formula is now intuitively very clear. The i-th epicycle contributes to the mean motion only when it is close enough to the origin so as to encircle it . The proportion of time t is that close is, on the average, W_i . Here W_i is the probability that $(n - 1)$ segments $|a_1|, \cdots |a_{i-1}|, |a_{i+1}|, \cdots |a_n|$ linked with random angles have length less than $|a_i|$.

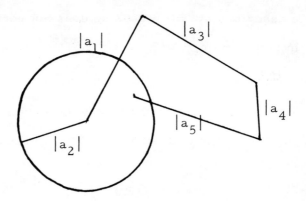

Notice that in computing $\int \Phi \, d\theta$ no special assumptions about the λ_i were used. If we take $\lambda_i = 1$ we get the amusing fact in probability

$$1 = W_1 + \dots + W_n \quad .$$

An explicit formula for W_i in terms of Bessel functions can be found in [W 1].

We now present Weyl's proof of the Kronecker-Weyl theorem. This theorem, together with many other fascinating number theoretical results can be found in Weyl's paper [W 5]. The theorem was first proved precisely to treat the problem of mean motion, cf. also [W 4].

LEMMA 1. <u>It is sufficient to prove the theorem for</u>
<u>continuous functions</u>.

Proof: Any bounded Riemann integrable function can
be approximated from above and below (by step functions
and therefore) by continuous functions in the sense that
given f we can find, for any $\epsilon > 0$, continuous g and
h such that

$$g \leq f \leq h \qquad \text{for all } x$$

and $h - g < \epsilon$ for all x except for $x \in A$ where $\mu(A) < \epsilon$.
Then $\frac{1}{T} \int_0^T f(\lambda t + \alpha)dt$ is caught between the corresponding expression for g and h. Since $\int (g - h)d\theta < \epsilon + M\epsilon$
where $M = \max_x(|g| + |h|)$ we have proved the lemma.

LEMMA 2. <u>It is sufficient to show that</u>

$$\int e(k \cdot \lambda t)dt = o(T),$$

<u>for</u> $k \neq 0$.

In fact, for any continuous f, we can for any ϵ,
find a trigonometric polynomial $g(\theta) = \Sigma a_\alpha e(\alpha \cdot \theta)$ such
that $|g(\theta) - f(\theta)| < \epsilon$ for all θ. Thus

$$\left| \frac{1}{T} \int_0^T f(\lambda t + \alpha) - g(\lambda t + \alpha) \right| < \epsilon$$

while $\frac{1}{T} \int g(\lambda t + \alpha) - a_0 \to 0$ by assumption, and $a_0 = \int g(\theta)d(\theta)$.

To show that $\int_0^T e(k \cdot \lambda t)dt = o(t)$ is just a computation:

$$\int_0^T e(k \cdot \lambda t + \alpha) = \frac{1}{2\pi i\, k \cdot \lambda} (e(k \cdot \lambda T) - 1)e(k \cdot \alpha)$$

which is bounded. (It is well defined because $k \cdot \lambda \neq 0$ by assumption.) This proves the theorem.

§3. <u>Weyl's argument</u>. As we have already pointed out, we can not apply the Kronecker-Weyl theorem to the function

$$\mathrm{Re}\ \frac{\Sigma\, \lambda_i e_i(\theta)}{\Sigma\, a_i\, e_i(\theta)}$$

since it is not bounded. Weyl avoids this by applying the following trick. Let S be the set of those θ with $z(\theta)$ real and negative, thus

$$S = \{\theta \,|\, \mathrm{im}\ \theta = 0 , \ \mathrm{Re}\ \theta < 0\} \quad .$$

$\{$The torus$\}$ - $\overline{S}$ is mapped by z into the complex plane, slit along the negative real axis, $\mathbb{C} - \mathbb{R}^-$ where $\mathbb{R}^-$ is the nonpositive real axis. On $\mathbb{C} - \mathbb{R}^-$ we can define the argument ϕ is an unambiguous manner so that it varies from $-\frac{1}{2}$ to $+\frac{1}{2}$, say

Thus φ becomes a well defined function on $\{$the torus$\}$ - $\overline{S}$. Now let $\theta(t) = \lambda \cdot t + \alpha$ be any straight line which does not pass through the singular set $\partial S = z^{-1}(0)$. Then for any choice of argument $\varphi(t)$ so that

$$z(\theta(t)) = r(t)e(\varphi(t))$$

it is clear that

$$\varphi(t_1) - \varphi(t_2) = N(t_1, t_2) + O(1)$$

where $N(t_1, t_2)$ is the number of times that $\theta(t)$ crosses
S . Here we count the crossings algebraically , +1 if we
go from $\phi = \frac{1}{2}$ to $\phi = -\frac{1}{2}$ and -1 otherwise. The term
O(1) is in fact always ≤ 1 . The function N is well defined
since $\theta(t)$ can not be tangent to S to infinite order. In
fact this would imply that

$$\Sigma \, a_k \, \sin(\lambda_k t + \alpha_k)$$

had a zero of infinite order which is impossible since the
λ_k's are distinct and the $a_k' \ne 0$. (In fact, it can't have
a zero of degree greater than 2n , because a zero of
order $2n + 1$ at t_0 implies equations of the form

$$\Sigma \, a_k \, \sin(\lambda_k t_0 + \alpha_k) = 0 = \Sigma \, \lambda_k a_k \, \cos(\lambda_k t_0 + \alpha_k)$$

$$\Sigma \, \lambda_k^2 a_k \sin(\lambda_k t_0 + \alpha_k) = 0$$

$$\vdots$$

$$\Sigma \, \lambda_k^{2n} a_k \sin(\lambda_k t_0 + \alpha_k) = 0 = \Sigma \, \lambda_k^{2n+1} a_k \cos(\lambda_k t_0 + \alpha_k) \quad .$$

Since $\lambda_k \ne \pm \lambda_i$ for any i and k we know that the
van-der Monde determinant $\Pi(\lambda_i^2 - \lambda_j^2) \ne 0$ implying that
$a_k \sin(\lambda_k t_0 + \alpha_k) = a_k \cos(\lambda_k t_0 + \alpha_k) = 0$, i.e., $a_k = 0$
for all k .)

Let us define the function $\eta(\theta)$ as

$$\eta(\theta) = \text{number of times } \theta - \lambda t \text{ crosses } S$$
$$\text{for } -1 \leq t \leq 0 .$$

Notice that $\eta(\theta)$ is well defined for all θ such that the segment $\theta - \lambda t$ does not intersect $\partial \overline{S} = z^{-1}(0)$. Since $z^{-1}(0)$ is a submanifold of dimension $n - 2$, the set where $\eta(\theta)$ is not defined has content zero. Furthermore

$$|\eta(\theta)| \leq M(\theta)$$

where $M(\theta)$ is the number of times that $\text{Re } z(\theta - \lambda t) = 0$ for $-1 \leq \lambda \leq 0$ (each crossing counted as ± 1). But $M(\theta)$ is bounded. In fact, $M(\theta) < \infty$ for any θ since $\Sigma a_k \sin(\lambda_k t + \alpha_n)$ has no zero of order $> 2n$. But then $|M(\theta - M(\theta_0)| < 4n$ for θ close enough to θ_0. Thus compactness guarantees that $M(\theta)$ and hence $\eta(\theta)$ is bounded. Furthermore the discontinuities of $\eta(\theta)$ are along the hypersurfaces $\theta \in \overline{S}$ and $\theta - \lambda \in \overline{S}$. Thus

$\underline{\eta(\theta) \text{ is a bounded Riemann integrable function.}}$

We can think of $\eta(\theta)$ in the following way: at each point of S draw the vector $\lambda = (\lambda_1, \cdots, \lambda_n)$.

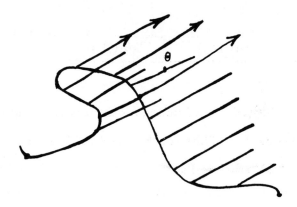

Then

$$\mathfrak{N}(\theta) \;=\; \Sigma \; \epsilon_p$$

where p ranges over those points of S for which the

vector λ covers θ , and $\epsilon_p = \pm 1$ according as the

crossing is in the positive or negative direction. We can

thus think of $\mathfrak{N}(\theta)$ as the "characteristic function" of the

"cylinder" erected over S by λ , where we count algebraic

multiplicity of how many times a point θ is covered by the

cylinder.

Now every time $\theta(t) = \lambda t + \alpha$ crosses S it will

stay in the cylinder for the following unit of time. That is,

$\mathfrak{N}(\theta(t))$ changes by ϵ_p for one unit of time past each

crossing. Thus up to a (bounded) contribution coming from

each end point,

$$N(t_1, t_2) = \int_{t_1}^{t_2} \mathcal{N}(\theta(t)) \, dt \quad .$$

Now $\mathcal{N}(\theta(t))$ is a bounded step function and therefore, Rieman integrable. Thus the Kronecker-Weyl theorem applies to $\mathcal{N}(\theta)$ and we have

$$\lim_{t_2 - t_2 \to \infty} \frac{\varphi(t_2) - \varphi(t_1)}{t_2 - t_1} = \int \mathcal{N}(\theta) \, d\theta \quad .$$

In view of the "cylinder" interpretation of $\mathcal{N}(\theta)$ we can compute this last integral as

$$\int \mathcal{N}(\theta) \, d\theta = \int_S \epsilon_p \lambda \lrcorner (d\theta_1 \wedge \cdots \wedge d\theta_n) \quad ,$$

i.e., as the "flux" of the vector field $\epsilon_p \lambda$ through S. That is,

$$\int \mathcal{N}(\theta) \, d\theta =$$

$$= \sum \epsilon_p (-1)^i \lambda^i \int_S \epsilon_p \, d\theta^1 \wedge \cdots \wedge d\theta^{i-1} \wedge d\theta^{S_i+1} \wedge \ldots \wedge d\theta^n \quad .$$

Now computing

$$\int_S \epsilon_p \, d\theta^1 \wedge \cdots \wedge d\theta^{n-1}$$

say, involves counting how many times

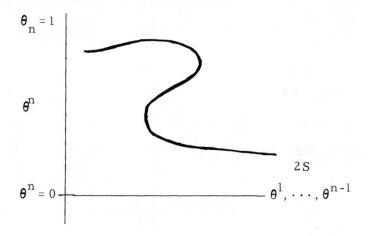

the surface S covers $\theta^1, \cdots, \theta^{n-1}$ (again with multiplicity included). Now let us hold $\theta^1, \cdots, \theta^{n-1}$ fixed and consider the circle

$$\sum_{k}^{n-1} a_k\, e_k(\theta_0) + a_n e_n(\theta) = A + a_n e_n(\theta) \quad .$$

If $|A| > a_n$ then either this circle will not cross the negative real axis or cross it twice - once in each direction.

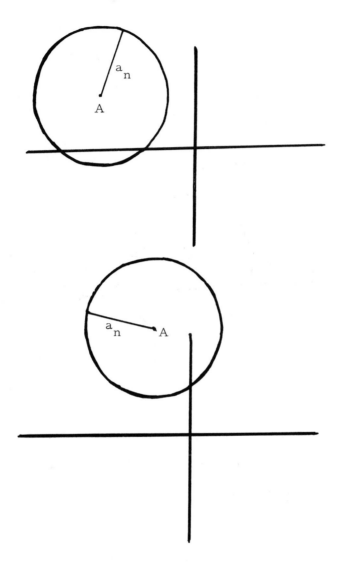

Thus the number of points of S over $(\theta_0^1, \cdots, \theta_0^{n-1}, 0)$ add up to zero. If $|A| < a_n$ then there will be exactly one crossing in the positive sense.

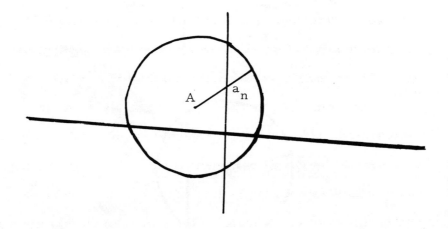

Thus the coefficient of λ_n is W_n , the measure of the set of those angles $\theta^1, \ldots, \theta^{n-1}$ for which $|A| < a_n$, again proving the formula.

§4. The ergodic theorem. The Kronecker-Weyl

theorem is the earliest example of an "ergodic theorem"

which allows one to replace a "time average" by a "space

average". This developed into a whole subject called

"ergodic theory" of which there are now many good exposit-

ions. The principle result is the celebrated Birkhoff

ergodic theorem. At the time of its proof this theorem

represented a tour-de-force of deep mathematical analysis.

It is a tribute to the development of the subject that by now

enough machinery has been developed to render the proof

of this theorem quite simple. The statement of the theorem

is:

Birkhoff ergodic theorem. Let T be a measure

preserving transformation of a finite measure space S .

That is $T^{-1}E$ is measurable whenever E is , and

$$\mu(T^{-1}E) \;=\; \mu(E)$$

for all measurable sets E . Then if f is an integrable

function then the limit

$$\hat{f}(p) \;=\; \lim_{n \to \infty} \; \frac{1}{n} \sum_{k=0}^{n=1} f(T^{n}p)$$

exists for almost all p, and $\hat{f}$ is an invariant integrable function such that

$$\int \hat{f} = \int f \quad .$$

For the elements of measure theory that we shall be using we refer to any standard book on real variables. To say that $\hat{f}$ is invariant means that $\hat{f}(Tp) = \hat{f}(p)$ which is clear from the definition of $\hat{f}$.

We say that the transformation T is _ergodic_ if any invariant set has measure zero, or its complement does: i.e., T is ergodic if

$$T^{-1}A = A \implies \mu(A) = 0 \quad \text{or} \quad \mu(A^i) = 0 \quad .$$

For any invariant function $\hat{f}$, the set $A_a = \{p \,|\, \hat{f}(p) < a\}$ is clearly invariant. Thus if T is ergodic the only invariant functions $\hat{f}$ are those which are constant (almost everywhere) . Thus for an ergodic T , the Birkhoff ergodic theorem asserts that the "time average"

$$\lim \frac{1}{n} \sum f(T^n p)$$

exists and equals the "space average" $\frac{1}{\mu(S)} \int f$ for almost all p . (Here we use a "discrete time" n . Passage to a continuous time, i.e., a one parameter group T_t , involves just a technical modification of our proof and will be omitted.)

Before proving the Birkhoff ergodic theorem let us first prove the so called "mean ergodic theorem" which was first proved by von-Neuman and which motivated G. D. Birkhoff to prove his theorem. Since T is measure preserving, $f \to T^*f = f(T \cdot)$ carries L^2 functions into L^2 functions and $(T^*f, T^*g^*) = \int T^*f \cdot \overline{T^*g} = \int f(Tp) \cdot \overline{g(Tp)}d\mu = \int f\bar{g}$ and so induces an isometry on L^2. Then we assert

The mean ergodic theorem. For any $f \in L^2$ there is an $\hat{f} \in L^2$ such that

$$\frac{1}{n} \sum T^{*k} f \longrightarrow \hat{f}$$

in the Hilbert space sense.

Proof: The limit, if it exists is certainly invariant. Thus we must prove that the limit exists for all $f \in L^2$. By this point, all that is relevant is that $T^* = U$ is a unitary operator on a Hilbert space H. If H were one dimensional then U is multiplication by $e^{i\alpha}$ and then

$$\frac{1}{n}(1 + e^{i\alpha} + \cdots + e^{in\alpha}) = \frac{1}{n} \frac{e^{i(n+1)\alpha} - 1}{e^{i\alpha} - 1} \quad \text{if } e^{in\alpha} \neq 1$$

and so the $\lim = 0$ and

and if $e^{i\alpha} = 1$ the expression is

$$\frac{1}{n}(1 + \cdots + 1) = 1$$

is identically one .

If H were a finite dimensional Hilbert space we could diagonalize U and conclude the theorem since all the eigenvalues of U are of absolute value one. For an infinite dimensional Hilbert space we could apply the spectral theorem and this was von-Neuman's original proof.

Actually, we can reset the above argument in a general Hilbert space and proceed as follows:

i) the set of all f for which the limit exists is a closed subspace of H: If

$$\frac{1}{2} \sum_{0}^{n-1} U^k g_i \longrightarrow \hat{g}_i , \qquad \text{and} \quad \| g_i - g_j \| < \epsilon$$

then

$$\left\| \frac{1}{n} \sum U^h g_i - \frac{1}{n} \sum U^k g_j \right\| < \epsilon \quad \text{so} \quad \| \hat{g}_i - \hat{g}_j \| \le \epsilon .$$

Then if $g_i \to f$ we conclude that $\hat{g}_i$ converges to some $\hat{f}$ and if $\| g_i - f \| < \epsilon$ we get

$$\left\| \frac{1}{n} \sum U^k f - \hat{f} \right\| \le \left\| \frac{1}{n} \sum U^k (g_i - f) \right\| + \left\| \sum U^k g_i - \hat{g}_i \right\|$$

$$+ \| \hat{g}_i - \hat{f} \| \le 3\epsilon \qquad \text{proving i).}$$

We now establish

ii) the set, E , of functions for which the limit exists is dense. For this it suffices to show that no $f \in H$ is orthogonal to E . Now for any $f \in H$ let $g = Uf - f$. Then

$$\lim_{n \to \infty} \frac{1}{n} \sum_{1}^{n} U^{k} g = \lim_{n \to \infty} \frac{1}{n} (U^{n+1} f - f) = 0$$

so $g \in E$. Now suppose $(f, g) = 0$. Then

$$0 = (f, \ Uf - f)$$

and

$$(Uf, Uf - f) = (Uf, Uf) - (Uf, f)$$

$$= (f, f) - (Uf, f) = (f - Uf, f) = 0 \ .$$

Therefore $(Uf - f, Uf - f) = 0$ so $Uf = f$. But then f is invariant so the limit exists trivially. This proves ii) and i) and ii) together gives the theorem .

Notice that we can replace L^2 by L^1 in the statement of the mean ergodic theorem. In fact, since S is of bounded measure, $L^2 \subset L^1$ since for $f \in L^2$

$$\|f\|_1^2 = (\int |f|)^2 \leq \|f\|_2 \cdot \|1\|_2 \ .$$

The proof of i) shows that the set of $f \in L^1$ such that

$$\lim \frac{1}{n} \sum_{0}^{n-1} T^{*k} f \quad \text{exists} \quad (\text{in } L^1)$$

is closed. Now L^2 is dense in L^1 and the above inequality shows that

$$\left\| \frac{1}{n} \sum T^{*k} f - \hat{f} \right\|_1 \le c \left\| \frac{1}{n} \sum T^{*k} f - \hat{f} \right\|_2^{1/2} \longrightarrow 0 .$$

Thus

$$\frac{1}{n} \sum T^{*k} f \quad \text{converges in } L^1$$

for all $f \in L^1$.

Let us now prove the Birkhoff ergodic theorem. For $f \in L^1$ we already have the candidate $\hat{f}$ which represents the limit - namely $\hat{f}$ must be the limit of $\frac{1}{n} \sum^{n} T^{*k} f$ in L^1 . We must show that this sequence of functions converges almost everywhere. Let B be the set of functions f for which $\frac{1}{n} \sum T^{*k} f(p) \to \hat{f}(p)$ almost everywhere. Notice that B contains all invariant f and all f of the form $f = T^* g - g$ with g bounded (by the same telescope argument as before, applied pointwise) . Now the closure of the set of both kinds of f is dense in

L^1 . We must show that B is closed in L^1 .

So suppose $f_j \in B$ and $\|f_j - f\|_1 \to 0$. Then for each p we have

$$\left| \frac{1}{n} \sum_0^{m-1} T^{*k}f(p) - \frac{1}{n} \sum_0^{n-1} T^{*k}f(p) \right| \le \left| \frac{1}{n} \sum_0^{m-1} T^{*k}f_j(p) - \frac{1}{n} \sum_0^{n-1} T^{*k}f_j(p) \right|$$

(*)

$$+ \ 2 \max \left| \frac{1}{m} \sum_0^{m-1} T^{*k}f(p) - T^{*k}f_j(p) \right|, \left| \frac{1}{n} \sum_0^{n-1} T^{*k}f(p) - T^{*k}f_j(p) \right| .$$

We would like to estimate the second term on the right since the first goes to zero for m, n, large. It would certainly be enough to prove

The maximal ergodic theorem(Hopf). Let $g \in L^1$.

Let

$$A = \left\{ p \ \middle| \ \sup_n \frac{1}{n} \sum_0^{n-1} T^{*k}g(p) > 0 \right\} .$$

Then

$$\int_A g \ge 0 .$$

In fact, assuming the maximal ergodic theorem, we get, taking $g = (f - f_j) - \lambda$ we see that on the set A_λ of p where

$$\sup_{m} \frac{1}{m} \sum T^{*k}f(p) - T^{*k}f_j(p) > \lambda$$

we have

$$\int_{A_\lambda} g \geq 0$$

or

$$\int f - f_j \geq \lambda \, \mu(A_\lambda) \quad .$$

Thus

$$\mu(A_\lambda) \leq \frac{1}{|\lambda|} \| f - f_j \|_1 \quad .$$

Then (*) shows that

$$\lim_{m, n \to \infty} \left| \frac{1}{m} \sum^{m-1} T^{*k}f(p) - \frac{1}{n} \sum^{n-1} T^{*k}f(p) \right| \leq 2|\lambda|$$

except on a set of measure at most $\frac{1}{|\lambda|} \| f - f_j \|$. Let $j \to \infty$ we see that the limit is $\leq |\lambda|$ except on a set of measure zero. Letting $\lambda \to 0$ proves the theorem. Thus we must prove the maximal ergodic theorem:

Let $A_\nu = \left\{ p \,\middle|\, \frac{1}{n} \Sigma_0^{n-1} T^{*k}f(p) \geq 0 \text{ for some } 0 \leq n \leq \nu \right\}$. Then $A_\nu \subset A_{\nu+1} \cdots$ and $\cup A_k = A$. We therefore have

$$\lim_{\nu \to \infty} \int_{A_\nu} g = \int_A g \quad .$$

So it suffices to prove

$$(\text{**}) \qquad \int_{A_\nu} g \geq 0 \, .$$

The proof which follows is due to Garsia [G 2] .

Notice that A_ν is the set where

$$g(p) + T^*g(p) + \cdots + T^{*n}g(p) > 0 \qquad \text{for some } 0 \leq n < \nu .$$

Let

$$h(p) = \max_{0 \leq n \leq \nu} (\{g(p) + T^*g(p) + \cdots + T^{*n}g(p)\}, 0)$$

so that

$$A_\nu = \{p \,|\, h(p) > 0\} \, .$$

Now

$$g + T^*h \geq g + T^*(g + T^*g + \cdots + T^{*n}g)$$

$$= g + T^*g + \cdots + T^{*n+1}g \quad \text{for any } 0 \leq n < \nu .$$

Now if $p \in A_\nu$ then $g(p) + T^*g(p) + \cdots + T^{*n+1}g(p) > 0$ for some $n \leq \nu - 1$ and so for suitable n we get h on the right. Thus

$$g(p) \geq h(p) - T^*h(p) \qquad \text{for } p \in A_\nu \, .$$

Now

64.

$$\int\limits_{A_\nu} g \geq \int\limits_{A_\nu} (h - T^*h) = \int h - \int\limits_{A_\nu} T^*h$$

since A_ν is precisely the set where $h \neq 0$. But then

$$\int\limits_{A_\nu} g \geq \int h - \int\limits_{A_\nu} T^*h \geq \int h - \int T^*h = 0 \quad .$$

This completes the proof of the mean ergodic theorem and therefore of the Birkhoff ergodic theorem .

§5. Examples of ergodic transformations. The ergodic theorem allows us to replace time averages (almost everywhere) by space averages for an ergodic transformation or flow. An ergodic transformation can be thought of as a transformation which moves any set of positive measure all around the space S in such a way that the relative amount of time any point spends in a given set is proportional to the size of the set. Thus the irrational straight line flow on the torus is ergodic. Notice that each small region of the torus is moved all around the torus without changing its shape.

Another type of ergodic transformation or flow is one which disperses any set of positive measure (< 1) all over S . In technical terms, a transformation T is called <u>mixing</u> if for any pair of measurable sets A and B, we have

$$\lim_{n \to \infty} \mu(T^{-n}A \cap B) \longrightarrow \mu(A) \cap \mu(B) \quad .$$

Notice that any mixing T is ergodic. In fact, if A is any invariant set we conclude that

$$\mu(A) = \mu(T^{n}A \cap A) \longrightarrow \mu(A)^{2}$$

so $\mu(A) = 0$ or one .

As we indicated above, the irrational flow on the torus is ergodic but not mixing. It turns out that one can establish sufficient criteria for a large class of dynamical systems to be mixing (Anosov's theorem), [A2] .

§6. Almost period functions and compact groups.

The concept of almost periodic functions of a real variable was generalized by von-Neuman and von Kampen to the case of a general group. Then A. Weil observed that, in fact, the study of almost periodic functions on a group can be reduced to the study of compact groups. Thus the Bohr approximation theorem becomes a consequence of the celebrated Peter-Weyl theorem on representations of compact groups. We will follow this approach here. An "elementary" proof of the Bohr approximation theorem can be found in Maak, "Fastperiodschen Funktionen."

Recall that a function f on $\mathbb{R}^1$ is almost periodic if the set of functions $f(\cdot + h)$ is sequentially pre-compact. Now to say that a family of functions is sequentially pre-compact is the same as saying that it is totally bounded in the uniform norm. (See Lemmas 5.2 and 7.4 in Chapter 4 of [L 3].) This suggests the following definition (due to von-Neuman) of almost periodicity on a group.

DEFINITION 6.1. A function f on a group G is almost periodic if the family of functions f_s on $G \times G$ is totally bounded in the uniform norm, where

$$(6.1) \qquad f_s(g, h) = f(gsh) \ .$$

We now observe that we can use f to introduce a notion of distance on G, namely define

$$(6.2) \qquad d_f(s, t) = \| f_s - f_t \|_\infty$$

where $\| \ \|_\infty$ denotes the uniform norm on $G \times G$. Thus

$$(6.3) \qquad d_f(s, t) = \sup_{g, h} |f(gsh) - f(gth)| \ .$$

It follows immediately from (6.3) that d_f is both right and left invariant, i. e. , that

$$d_f(s, t) = d_f(su, tu) = d_f(us, ut)$$

for any $u \in G$.

Let H be the set of $s \in G$ satisfying

$$d_f(s, e) = 0$$

where e is the identity element of the group. Then if s and t lie in H

$$d_f(st, e) \leq d_f(st, t) + d_f(t, e) = d_f(s, e) + d_f(t, e) = 0$$

and

$$d_f(utu^{-1}, e) = d_f(t, u^{-1}eu) = d_f(t, e) = 0$$

so that H is a normal subgroup . Thus d_f induces a metric on the quotient group G/H .

Notice that

$$d_f(s_1 t_1, s_2 t_2) \leq d_f(s_1 t_1, s_2 t_1) + d_f(s_2 t_1, s_2 t_2)$$

$$= d_f(s_1, s_2) + d_f(t_1, t_2)$$

and

$$d_f(s^{-1}, t^{-1}) = d_f(s s^{-1}t, s t^{-1}t) = d_f(s, t)$$

so that multiplication on G/H is continuous relative to

68.

this metric. Since the set of all functions f_s is totally
bounded we can find, for any $\epsilon > 0$, a finite number of
s_i such that

$$\min_i \, d_f(s, s_i) < \epsilon$$

for any $s \in G$. Thus G/H is a totally bounded metric
space. Finally, f is constant on cosets of H since

$$\left| f(s) - f(sh) \right| = \left| f(s\,e\,e) - f(s\,h\,e) \right| \leq d_f(h, e) = 0 .$$

It thus induces a continuous function $\hat{f}$ on G/H. We
have thus proved

THEOREM 6.1. (A. Weil). An almost periodic
function f on a group G induces a homomorphism
$G \to G/H$ of G into a group with a bi-invariant metric, d_f.
The group G/H is a totally bounded topological group
under this metric and f induces a continuous functions, $\hat{f}$,
on G/H.

Of course, we could complete G/H to get a compact
group, but this will be un-necessary for what follows.

§7. The mean ergodic theorem and the Haar
integral. In this section we shall establish the existence
of a bi-invariant mean or integral for uniformly continuous

functions on a totally bounded group. By applying Theorem 6.1 we will obtain the existence of a mean for almost periodic functions. We shall establish the existence of the mean by applying a version of the mean ergodic theorem, an idea due to Yosida [Y1] .

We first observe that we can reformulate the mean ergodic theorem as follows:

THEOREM 7.1. <u>Let</u> T <u>be a linear transformation of a normed linear space such that</u>

(i) $\|T^n x\| \leq c \|x\|$ <u>for all</u> $x \in B$ (c <u>independent of n</u>).

(ii) <u>For some fixed</u> $y \in B$ <u>the sequence</u>

$$S_n y = \frac{1}{n}(y + Ty + \cdots + T^{n-1} y)$$

<u>possesses a convergent subsequence , converging to</u> $\overline{y}$.

<u>Then</u> $T\overline{y} = \overline{y}$ <u>and the sequence</u> $S_n y$ <u>converges to</u> $\overline{y}$.

We first observe that (i) implies that $\overline{(I - T)B}$ consists of all $x \in B$ such that $S_n x \to 0$. In fact, if $x = (I - T)z$ then

$$\|S_n x\| = \left\|\frac{1}{n}(T^n z - z)\right\| \leq \frac{2c}{n}\|z\| \longrightarrow 0 \ ,$$

so that

$$(I - T)B \subset \left\{x : S_n x \longrightarrow 0\right\}$$

and the right side is closed by (i) . If $S_n x \to 0$ then

$$\| x - (x - S_n x) \| < \epsilon \qquad \text{for large } n .$$

But for any n

$$x - S_n x \in (I - T)B .$$

In fact

$$x - S_n x = \frac{1}{n}\left\{(I - T)x + (I - T^2)x + \cdots + (I - T^{n-1})x\right\}$$

$$= \frac{1}{n}\left\{(I - T)x + (I - T)(I + T)x + \cdots + (I - T)(I + \cdots + T^{n-2})x\right\}$$

so $x - S_n x \in (I - T)B$. Thus $S_n x \to 0 \implies x \in \overline{(I - T)B}$.

If $S_{n_j} y \to \overline{y}$ then

$$TS_{n_j} y - S_{n_j} y = \frac{1}{n}(T^{n+1} y - y) \longrightarrow 0$$

so

$$T\overline{y} = \overline{y} .$$

Now $T^n y = T^n \overline{y} + T^n(y - \overline{y}) = \overline{y} + T^n(y - \overline{y})$

so

$$S_n y = \overline{y} + S_n(y - \overline{y}) .$$

By assumption

$$y - S_{n_j} y \longrightarrow y - \overline{y}$$

and

$$y - S_{n_j} y \in (I - T)B .$$

Thus $y - \overline{y} \in \overline{(I - T)B}$ so that

$$S_n(y - \overline{y}) \longrightarrow 0 \ .$$

(Remark. We really only needed to assume that $S_{n_j} y \to \overline{y}$

weakly, i.e., that for continuous linear functionals f

$$f(S_{n_j} y) \longrightarrow f(\overline{y}) \ .$$

In fact, as before we conclude that $0 = f(T\overline{y} - \overline{y})$

$= \lim f(TS_{n_j} y - S_{n_j} y)$. Since this holds for all f we have

$T\overline{y} = \overline{y}$. Also, it follows that $y - \overline{y} \in \overline{(I - T)B}$. In fact,

if $y - \overline{y} \notin \overline{(I - T)B}$ then, by the Hahn-Banach theorem,

there will be a functional f vanishing on $\overline{(I - T)B}$ with

$f(y - \overline{y}) \neq 0$. But $y - S_{n_j} y \in (I - T)B$ and $f(y - S_{n_j} y)$

$\to f(y - \overline{y})$ yielding a contradiction. This then reproves the

mean ergodic theorem in Hilbert space since the unit ball

is weakly compact.)

We will now apply the theorem to totally bounded

groups. Let K be such a totally bounded group. This

means that there is a fundamental family of neighborhoods,

U_i, of e , and for each U_i a finite number of elements

$g_1, \cdots, g_{j(i)}$ such that the sets $g_1 U_i, \cdots, g_{j(i)} U_i$ cover M.

(Thus if the topology comes from a bi-invariant metric d_f

then we could take U_i to be balls of radius $1/i$ about the

origin, if K is totally bounded in the metric sense. Also,

if K is compact with a countable basis of neighborhoods, U_i of e we could find the g_j.)

Take all the g's and collect them into a sequence $\{k_i\}$. Thus $\{k_i\}$ is dense in K.

We let B be the space of uniformly continuous functions on K with the sup norm, i.e.,

$$\|f\| = \sup_{k \in K} |f(k)| \quad .$$

(If K is compact every continuous function is uniformly continuous.)

Define the operator T on B by

$$(Tf)(k) = \sum \frac{1}{2^j} f(k_j^{-1}k) \quad .$$

If $|f(u) - f(v)| < \epsilon$ whenever $u^{-1}v \in U$ where U is some neighborhood of e then

$$|Tf(u) - Tf(v)| \leq \sum \frac{1}{2^j} |f(k_j^{-1}u) - f(k_j^{-1}v)| < \epsilon$$

so T maps B into itself. It is clear that $\|T\| \leq 1$.

Now

$$S_n f = \frac{1}{n} \sum_0^{n-1} T^k f$$

has the form

$$S_n f(k) = \sum_i a_i f(g_i k) , \qquad s_i \in k$$

where $\Sigma a_j = 1$. By the above argument, $S_n f$ is equi-continuous, i. e.,

$$u^{-1} v \in U \implies \left| S_n f(u) - S_n f(v) \right| < \epsilon .$$

Since K has a dense sequence, this implies that we can choose a convergent subsequence of $S_n f$. Thus we can take the y in Theorem 7.1 to be any $f \in B$, and conclude that

$$S_n f \longrightarrow \overline{f} \quad \text{where} \quad T\overline{f} = \overline{f} .$$

We claim that $\overline{f}$ is a constant. By taking real and imaginary parts we may assume f is real valued. (If f is vector valued first apply a linear functional). Let $m = \sup \overline{f}$. If $\overline{f}$ is not constant then there will be some ℓ with $\overline{f}(\ell) < m - 2\epsilon$ for some $\epsilon > 0$. Then since $\overline{f}$ is uniformly continuous

$$\overline{f}(k) < m - \epsilon$$

if $k\ell^{-1} \in U$ where U is some suitable neighborhood of e. But by construction, a finite number of the $k_i U$ cover all of K . Thus for any $g \in K$ we will have

$$g \, \ell^{-1} \in k_i U \qquad \text{for some } i < N$$

and thus

$$(k_i^{-1} g)\ell^{-1} \in U$$

or

$$\overline{f}(k_i^{-1} g) < m - \epsilon \ .$$

But then

$$T\overline{f}(g) = \sum \frac{1}{2^j} \overline{f}(k_j^{-1} g) \leq \sum_{j \neq i} \frac{1}{2^j} m + \frac{m - \epsilon}{2^i}$$

$$\leq m - \frac{\epsilon}{2^N}$$

since $i < N$. Since g is arbitrary, this contradicts the definition of m. Thus $\overline{f}$ is constant. We shall denote it as $M(f)$.

We can state some properties of M.

(7.1) M is a linear functional

(7.2) $M(1) = 1$

(7.3) $M(h) \geq 0$ if $h(g) \geq 0$ for all   s.

(7.4) $M(h(\cdot g)) = M(h(\cdot))$.

These follow directly from the definitions. Notice that by definition there exists for any h and $\epsilon > 0$, $a_j \geq 0$

with $\Sigma\, a_j = 1$ and $g_j \in G$ such that

$$(*) \qquad \| \Sigma\, a_j\, h(g_j \cdot) - M(h) \| \le \epsilon \quad .$$

Now we could equally well have started with a $\hat{T}$ in the form $\hat{T} f(x) = \Sigma\, \dfrac{1}{2^k}\, f(xk_i)$ (replacing left by right multi-plication) and end up with an N satisfying (7.1), (7.2) and (7.3) with (7.4) replaced by

$$N(h(g \cdot)) = N(h) \quad .$$

For a fixed h and $\epsilon > 0$ we could find $b_j \ge 0$ with $\Sigma\, b_j = 1$ and ℓ_j such that

$$(**) \qquad \| \Sigma\, b_j\, h(\cdot\; \ell_j) - N(h) \| \le \epsilon \quad .$$

Then (*) and (7.4) implies that

$$\sup_g \left| \Sigma\, a_i\, b_j\, h(g_i g \ell_j) - M(h) \right| \le \epsilon$$

and similarly from (**)

$$\sup_g \left| \Sigma\, a_i\, b_j\, h(g_i\, g \ell_j) - N(h) \right| \le \epsilon \quad .$$

Thus $M(h) = N(h)$ so we have

$$(7.5) \qquad M(h(g \cdot)) = M(h) \quad .$$

Notice also that (7.1) - (7.4) uniquely determines M. In fact let $\overline{M}$ be any functional satisfying (7.1) - (7.4) . Since $N = M$ in (**) we have

$$M(h) - \epsilon \leq \Sigma b_j h(g \ell_j) \leq M(h) + \epsilon$$

for all $g \in K$ so that since $\overline{M}(\Sigma b_j h(g \ell_j)) = \overline{M}(h)$ by (7.4)

$$\left| M(h) - \overline{M}(h) \right| < \epsilon$$

i.e., $M = \overline{M}$.

We also remark that

(7.6) $\quad h \geq 0$ and $h(\ell) > 0$ for some $\ell \Rightarrow M(h) > 0$.

In fact, we may assume that

$$h(g) > \epsilon \quad \text{for} \quad \ell g^{-1} \in U$$

where U is some neighborhood of e . Choose finitely many g_i $(i = 1, \cdots, n)$ so that $g_i^{-1} U$ cover K . Then for any $s \in K$

$$g_i s \ell^{-1} \in U \qquad \text{for some} \quad 1 \leq i \leq n .$$

Then

$$\sum h(g_i s) > \epsilon$$

so

$$M(\Sigma h(g_i \cdot) = n M(h) > \epsilon \qquad \text{proving (7.5).}$$

We finally remark that (*) implies the usual Schwartz inequality:

$$(7.7) \qquad |M(f_1 \overline{f}_2)|^2 \leq M(|f_1|^2)M(|f_2|^2) \ .$$

By applying Theorem 6.1 and the previous result we have established,

THEOREM 7.2 (von Neuman). <u>Let</u> A <u>be the space of almost periodic functions on a group</u> G . <u>There exists a unique linear functional,</u> M, <u>on</u> A <u>satisfying (7.1) - (7.4). It satisfies, in addition, (7.5) and (7.6)</u> .

In fact, for any almost periodic function f, define $M(f) = M(\hat{f})$ where $\hat{f}$ is the uniformly continuous function on G/H given by Theorem 6.1 . From the definitions it follows that (*) and (**) still hold where now we regard the g_j as elements of G . This proves uniqueness as before. (To prove linearity we again use (*) where we choose the g_j to work for both f_1 and f_2 as we can clearly do.)

It is an instructive exercise to check directly that for almost periodic functions on $\mathbb{R}$ the limit

$$\lim_{T \to \infty} \frac{1}{T} \int_0^T f(t)\,dt$$

exists and satisfies $(7.1) - (7.4)$ so that

$$(7.8) \qquad M(f) = \lim_{T \to \infty} \frac{1}{T} \int_0^{\bar{\ }} f(t)dt \quad .$$

We can derive it from the Bohr approximation theorem (to be proved in § 9) . In fact, by (7.4) and (7.1),

$$M(e^{i\lambda \cdot}) = M(e^{i\lambda(\cdot + a)}) = e^{i\lambda}M(e^{i\lambda \cdot})$$

where $e^{i\lambda \cdot}$ is the function whose value at t is $e^{i\lambda t}$ so

$$M(e^{i\lambda \cdot}) \begin{array}{l} = 0 \\ = 1 \end{array} \quad \text{if} \quad \begin{array}{l} \lambda \neq 0 \\ \lambda = 0 \end{array}$$

and the same holds for the right hand side of (7.8) . Since, by the Bohr approximation theorem, the functions $e^{i\lambda \cdot}$ span a dense subspace B of the space of almost periodic functions. Thus (7.8) holds.

§ 8. The Peter-Weyl theorem. As before, let B denote the space of all uniformly continuous functions on a totally bounded group G .

DEFINITION 8.1. A function f is called a representative function if the space spanned by all $f_g = f(g \cdot)$ is finite dimensional. The set of all representative functions will be denoted by $\mathfrak{R}$.

Suppose that $f \in \Re$. Let $f_{g_1}, \cdots, f_{g_n}$ be a maximally linearly independent family of f_g's . Then for any g we have

$$(8.1) \qquad f(gu) = f_g(u) = \sum h_i(g) f_i(u)$$

where $f_i(u) = f_{g_i}(u)$ the h_i are some functions on G. Thus if $f \in \Re$ there are functions h_i and f_i such that (8.1) holds. Conversely, if (8.1) holds then it is obvious that $f \in \Re$. From (8.1) it is clear that if $f \in \Re$ then the functions $f(\cdot g)$ also span a finite dimensional space. It is also clear that if we set

$$\tilde{f}(x) = f(x^{-1})$$

then

$$(8.2) \qquad f \in \Re \implies \tilde{f} \in \Re .$$

The Peter-Weyl theorem asserts:

THEOREM 8.1. <u>The set $\Re$ is dense in</u> B .

To prove this we introduce and study the convolution operator $*$ defined by

$$f_1 * f_2(x) = M_g(f_1(g)f_2(g^{-1}x))$$

where M_g means that we apply M to the corresponding

function of y. The convolution is defined on $B \times B$. We define $L_1(G)$ to be the completion of B in the norm

$$\|f\|_1 = M(|f|)$$

and $L_2(G)$ to be the completion of B with respect to the norm

$$\|f\|_2 = M(|f|^2)^{1/2} .$$

By (7.7)

$$\|f\|_1 \leq M(|f|^2)^{1/2} M(1)^{1/2} = \|f\|_2$$

so, we have, for $f, h \in B$

$$|f * h(x)| \leq \|f\|_2 \|h\|_2$$

so that $*$ extends to a map from $L_2 \times L_2 \to B$.

LEMMA 8.1. <u>Let</u> $f \in B$. <u>Then the family of functions</u> $f * h$, $\|h\|_2 \leq 1$ <u>is uniformly equicontinuous.</u>

Proof: For any $\epsilon > 0$ choose a neighborhood U such that

$$|f(yx) - f(x)| < \epsilon \qquad \text{for all } y \in U .$$

Then

$$f * h(yx) - f * h(x) = M_z\{(f(yxz) - f(xz))h(z^{-1})\}$$
$$\leq \epsilon M(|h|) \leq \epsilon \|h\|_2 ,$$

proving the lemma.

LEMMA 8.2. If $f \in B$ then the map $h \rightsquigarrow f * h$ is a compact operator from $L_2 \to B$ and a fortiori a compact operator from $L_2 \to L_2$.

(For the definition and elementary properties of compact cf. [L3] p. 264.) For any sequence $\{h_\alpha\}$ with $\|h_\alpha\|_2 \leq 1$ we may choose a subsequence such that (h_2, φ) converges for any $\varphi \in L_2$, (since the unit ball in a Hilbert space is sequentially compact.) But then, for any $x \in G$,

$$(f * h_\nu)(x) = M_y(f(xy^{-1})h_\nu(y))$$

$$= (f(x \cdot {}^{-1}), h_\nu)_2$$

and so converges. Since $f * h_\nu$ is uniformly equicontinuous and G is totally bounded we can thus choose a convergent subsequence.

PROPOSITION 8.1. The space $\Re$ is dense in L_2.

Proof: Suppose $0 \neq f \in L_2$ is orthogonal to $\Re$. Then for any $h \in \Re$,

$$f * h(x) = M_y(f(y)h(y^{-1}(x)))$$

$$= M_y(f(y), \overline{\tilde{h}(x^{-1}y)})$$

$$= (f, h(x^{-1} \cdot))_2 = 0$$

so $f * h = 0$ for all $h \in \mathfrak{R}$. Now $\tilde{f} * f \in B$ so the operator A given by

$$Ah = \tilde{f} * f * h$$

is compact.

Now

$$(\tilde{f} * \varphi, \psi) = M_x(M_y((f(x^{-1}y)\varphi(y))\psi(x))$$

$$= (\varphi, f * \psi)$$

so that

$$(\tilde{f} * f * \varphi, \psi) = (f * \varphi, f * \psi)$$

and so A is self-adjoint. Furthermore, $A \neq 0$ since

$$\tilde{f} * f(e) = M_y(f(y) \cdot f(y)) = \|f\|^2$$

so taking $\varphi = \psi = \tilde{f}$ in the preceeding equation gives a non-zero result.

Since A is self-adjoint, compact, and nonzero, it must have a nonzero eigenvector λ cf.[L 3] p. 264 . Let W be a nonzero eigenspace of A with eigenvalue λ . Then W is finite dimensional. If $h \in W$ then

$$h = \frac{1}{\lambda} Ah = \frac{1}{\lambda} \tilde{f} * f * h$$

so that h is continuous. Also

$$\lambda h_z(x) = \lambda h(xz) = (\tilde{f} * f) * h(xz)$$

$$= ((\tilde{f} * f)(y)h^{-1}xz))$$

$$= (\tilde{f} * f * h_z)(x)$$

where $h_z(x) = h(xz)$. Thus $h_z \in W$, and h is a representative function. But then $Ah = \tilde{f} * f * h = 0$ contrasting $\lambda \neq 0$. Thus we must have $f = 0$, proving the proposition.

Proof of Theorem 8.1. Let $f \in B$ and choose a neighborhood U of e so that $|f(xy^{-1}) - f(x)| < \epsilon$ for $y \in U$. Let $\overline{\varphi} \geq 0$ be a nonzero function in B whose support lies in U . (If G has a metric, d, we simply take $\overline{\varphi} = \rho(d(\cdot, e))$ where ρ is a suitable real valued function. Otherwise the existence of $\overline{\varphi}$ is guaranteed by a standard theorem in point set topology (Urysohn's lemma). Let $\varphi = \dfrac{\overline{\varphi}}{M(\overline{\varphi})}$. Then

$$\left| f * \varphi(x) - f(x) \right| = \left| M_y(f(xy^{-1}))\varphi(y)) - f(x) \cdot M(\varphi) \right|$$

$$\leq \epsilon .$$

Now by Proposition 8.1 we can find an $h \in \Re$ such that

$$\|h - f\|_2 \leq \dfrac{\epsilon}{\|\varphi\|_2} .$$

Then

$$\left| f(x) - \varphi * h(x) \right| \le \epsilon + \left| \varphi * f(x) - \varphi * h(x) \right|$$

$$\le \epsilon + \left\| \varphi \right\|_2 \left\| f - h \right\|_2 \le 2\epsilon \quad .$$

This completes the proof of the Peter-Weyl theorem .

Let $x \to A_x$ be a representation of G on a finite dimensional vector space V such that the functions A_x are uniformly continuous. If we choose a basis for V we can realize A_x as a matrix $(a_{ij}(x))$. Since $x \to A_x$ is a representation we see that $a_{ij}(\cdot y)$ is a linear combination of the $a_{ij}(\cdot)$. Thus all matrix elements of finite dimensional representations belong to $\Re$. Conversely, if $f \in \Re$ then let V be the space spanned by the translates of f , and let $A_x h = h(\cdot x^{-1})$ for any $h \in V$. By $(\ ,\ )$

$$f(xy) = \Sigma\, h_i(x)\, f_i(y)$$

where $f_i = f(\cdot\, x_i)$ so

$$A_u f_j = f(\cdot\, x_j u^{-1}) = \Sigma\, h_i(x u^{-1})\, f_i$$

so

$$a_{ij}(u) = h_i(x_j u^{-1})$$

gives the metric representation of A_u with respect to the basis f_i . But then

$$f(u) = A_{u^{-1}} f(x_j x_j^{-1})$$

$$= A_{u^{-1}} f_j(x_j) = \Sigma\, a_{ij}(u^{-1})\, f_i(x_j)\ .$$

Thus f is a linear combination of the functions $a_{ij}(u^{-1})$ which are matrix elements for the representation $\alpha \rightsquigarrow A_{u^{-1}}^{t}$. Thus

PROPOSITION 8.2. <u>The space</u> $\Re$ <u>is the space of</u> <u>linear combinations of matrix elements of finite dimensional</u> <u>representations of</u> G .

This is why $\Re$ is called the space of representative functions.

If $x \rightarrow A_x$ is a finite dimensional representation of G on a vector space V , one can always introduce a scalar product on V so that all the A_x are unitary . In fact, choose an arbitrary scalar product $(\ ,\ )_0$ and define $(\ ,\ )$ by

$$(\xi, \eta) = M_x(A_x \xi, A_x \eta)\ .$$

It is easy to check that $(\ ,\ )$ is indeed a scalar product and that the A_x are unitary .

In particular, if V_1 is an invariant subspace then $V_1^{\perp}$ is an invariant compliment. We may therefore always

choose basis so that the a_{ij} reflect this decomposition .
In particular, we can improve on Proposition 8.2 by saying
that $\mathfrak{R}$ is the span of matrix elements of irreducible
representations.

§9. <u>The Bohr approximation theorem</u>. We refer
the reader to any standard book on group representations
for the many applications of the Peter-Weyl theorem .

For the case of almost periodic functions on $\mathbb{R}$,
the groups we wish to study are all abelian. For an abelian
group it is easy to check that an irreducible representation
is one dimensional. In fact if $x \to A_x$ is irreducible then
each A_x can have only one eigenvalue since the eigenspace
of A_x is an invariant subspace. Thus all A_x will be
multiples of the identity, i.e., the representation is one-
dimensional.

A <u>character</u> on an abelian group G is a continuous
function, χ with $|\chi(x)| = 1$ such that $\chi(ab) = \chi(a)\chi(b)$ for
any $a, b \in G$. The element of a one by one matrix (a_{11})
of a representation and conversely. Thus the Peter-Weyl
theorem asserts

PROPOSITION 9.1. <u>On any totally bounded group</u>
G <u>the linear combination of the characters are dense in</u> B.

<u>Proof of the Bohr approximation theorem:</u> Let f

be an almost periodic function on $\mathbb{R}$. Construct the group

$G = \mathbb{R}/H$ as in §6. Any character χ on G is a uniformly

continuous function of G and by the homomorphism $\mathbb{R} \to \mathbb{R}/H$,

gives a character of $\mathbb{R}$.

Now since, $\mathbb{R}$ is commutative, the map $\mathbb{R} \to \mathbb{R}/H$

is given by

$$a \rightsquigarrow f(\cdot + a)$$

so that

$$a \rightsquigarrow \chi(f(\cdot + a))$$

is the character of $\mathbb{R}$. But every character of $\mathbb{R}$ is

easily seen to be a function of the form $a \rightsquigarrow e^{i\lambda a}$. (In

fact if φ is a character on $\mathbb{R}$ then, setting $\varphi(r) = e^{i\theta}$ we

see that $\varphi(r/2) = \pm e^{i\frac{\theta}{2}}$. If $|r|$ is sufficiently small

$e^{i\theta}$ must be close to one and so must be $\varphi(r/2)$ so that

$\varphi(r/2) = +e^{i\frac{\theta}{2}}$. Similarly $\varphi(r/2^n) = e^{i\frac{\theta}{2^n}}$ and by continuity

we conclude that $\varphi(ra) = e^{i\theta a})$. Thus

$$\chi(f(\cdot + a)) = e^{i\lambda a} .$$

Now consider the continuous function $\hat{f}$ on G sending

$h \in G$ into $h(0)$. By the Peter-Weyl theorem we can find

for any $\epsilon > 0$, characters $\chi_1, \cdots, \chi_n$ such that

$$\left| h(0) - \Sigma\, c_j\, \chi_j(g) \right| < \epsilon .$$

Setting $h = f(\cdot + a)$ and $\chi_k(f(\cdot + a)) = e^{i\lambda_k a}$ yield

$$\left| f(a) - \Sigma\, c_k\, e^{i\lambda_k a} \right| < \epsilon$$

which is the Bohr approximation theorem.

Now the mean value M makes the space of almost periodic functions into a (non-separable) Hilbert space. As we have already observed the functions $e^{i\lambda \cdot}$ are orthonormal since

$$M_y\!\left(e^{i\lambda_1 y} e^{-i\lambda_2 y}\right) = \lim_{T \to \infty} \frac{1}{T} \int_0^T e^{i(\lambda_1 - \lambda_2)y} = \begin{array}{l} 0 \ \ \text{if } \lambda_1 \neq \lambda_2 \\ 1 \ \ \text{if } \lambda_1 = \lambda_2 \end{array}.$$

By the Bohr approximation theorem the linear span of the $e^{i\lambda_1}$ are dense in the uniform norm and therefore are certainly dense in the L_2 norm . Thus we can write

$$(9.1) \qquad f = \sum_1^\infty a(\lambda_\nu) e^{i\lambda_\nu \cdot} . \qquad \text{(convergence in the } L_2 \text{ sense)}$$

where each $a(\lambda_1)$ is given by

$$a(\lambda_\nu) = M(fe^{-i\lambda_\nu \cdot}) = \lim_{T \to \infty} \frac{1}{T} \int_0^T e^{-i\lambda_\nu t} f(t)\, dt \quad .$$

Also the λ_ν are determined by the condition that $M(fe^{-i\lambda_\nu \cdot}) \neq 0$. In particular, we have

$$(9.2) \qquad \|f\|_2^2 \; = \; M(f\bar{f}) \; = \; \Sigma \; |a(\lambda_\nu)|^2 \quad .$$

Those λ_ν that enter with non-zero coefficients in (9.1) are called the frequencies of f .

Of course, the "Fourier series" (9.1) will not, in general, converge. However, it is not difficult to show that if various summation methods are applied to (9.1) it will converge in the uniform norm. Let us show how to use (9.1) and (9.2) to construct approximations to f by trigonometric polynomials. We shall also see that it suffices to consider polynomials $P_n = \Sigma^n b_k e^{i\lambda_k}$ where λ_k are actual frequencies of f . In fact for any $\epsilon > 0$ let h be chosen so small that $\left| f(x+h) - f(x) \right| < \frac{\epsilon}{2}$ for all x and choose ℓ so large that each interval of length $\ell/2$ contains an $\epsilon/2$ almost period. Next choose N so large that

$$\sum_{\nu > N} |a(\lambda_\nu)|^2 \; < \; \frac{\epsilon^2}{2} \; \frac{h}{\ell + h} \quad ,$$

and set

$$f_N \; = \; f - \sum_{\nu \leq N} a(\lambda_\nu) e^{i\lambda_\nu} \quad .$$

Then setting $T = k(\ell + h)$,where k is any integer ,we conclude that the interval $(0, T)$ contains k nonoverlapping intervals

$(\tau_1, \tau_1 + h), \cdots (\tau_k, \tau_k + h)$ such that

$$|f(x + s) - f(x)| < \epsilon$$

when s lies in any of these intervals. Let e be the characteristic function of the union of these intervals so that $e(x) = 1$ or 0 according to whether x lies in one of the intervals or not. Then Schwartz's inequality yields

$$\left| \sum_1^k \int_0^h f_N(x + s + \tau_i)dx \right|^2 = \left| \int_0^T f_N(x + s)e(x)dx \right|^2$$

$$\leq kh \int_0^T \left| f_N(x + s) \right|^2 dx .$$

Now choose k so large that

$$\left| M(|f_N|)^2 - \frac{1}{T}\int_0^T |f_N(x + s)|^2 \right| < \frac{\epsilon^2 h}{2 (\ell + h)}$$

so that the right hand side of the previous inequality is less than $kh\, T^2 e^2 h/(\ell + h)$

$$\left| \frac{1}{kh} \sum_1^k \int_0^h f_N(x + s + \tau_i)dx \right| \leq \frac{1}{(kh)^{1/2}} T^{1/2}\left(\frac{h}{\ell+h}\right)^{1/2}\epsilon = e .$$

But

$$\int_0^h f_N(x + s + \tau_i) = \int_0^h f(x + s + \tau_i)$$

$$- \sum_1^N a(\lambda_\nu) e^{i\lambda_\nu(s+\tau_i)} \int_0^h e^{i\lambda_\nu x} dx$$

so

$$\frac{1}{kh} \sum_1^k \int_0^h f_N(x + s + \tau_i) dx = \frac{1}{kh} \sum \int_0^h f(x + s + \tau_i) dx - P_N(s)$$

where P_N is a trigonometric polynomial in s with frequencies $\lambda_1, \cdots, \lambda_N$.

But

$$\left| f(x + s + \tau_i) - f(s) \right| < \epsilon \qquad \text{for all } 0 \le x \le h$$

and thus

$$\left| \frac{1}{kh} \sum \int_0^h f(x + s + \tau_i) - f(s) \right| < \epsilon \ .$$

We thus get

$$\left| f(s) - P_N(s) \right| < 2\epsilon \ .$$

Notice that if $P_N = \Sigma \, b_\nu^N e^{i\lambda_\nu}$ and $P_N \to f$ uniformly then $\| P_N - f \|_2^2 \to 0$ so that

$$b_\nu^N \longrightarrow a(\lambda_\nu) \ .$$

We refer the reader to [W8] for a study of the
practical problems involved in actually determining the
frequencies λ_ν from empirical data. The methods are
based on the following simple consequence of the Bohr
approximation theorem:

PROPOSITION 9.2. <u>Let</u> f <u>be an almost periodic</u>
<u>function whose Fourier series is given by</u> (9.1) <u>and</u> λ <u>a</u>
<u>real number.</u> <u>Then the sequence</u>

$$\frac{1}{n}\left\{f(\cdot) + f(\cdot + \lambda) + \cdots + f(\cdot + (n - 1)\lambda)\right\}$$

<u>converges in the uniform norm to a function</u> f_λ <u>of period</u>
λ . <u>The (ordinary) Fourier series of</u> f_λ <u>is</u>

$$\sum_{\lambda_\nu = k\lambda} a(k\lambda)e^{i\lambda_\nu}\ ,$$

i.e., <u>the sum is over those frequencies which are multiples</u>
<u>of</u> λ .

The proposition is obvious for function of the form
$e^{i\mu \cdot}$ and thus for any trigonometric polynomial P_N .
Choose $P_N = \Sigma\, b_\nu^N\, e^{i\lambda_\nu \cdot}$ with $\|P_N - f\| < \epsilon$ in
the uniform norm. Then

$$\left\| \frac{1}{n}\{f + f(\cdot + \lambda) + \cdots + f(\cdot + (n-1)\lambda)\} \right.$$

$$\left. - \frac{1}{n}\{P_N + P_N(\cdot + \lambda) + \cdots + P_N(\cdot + (n-1)\lambda)\} \right\|$$

$$< \epsilon$$

and for $n \to \infty$, $\dfrac{1}{n}\{P_N + \cdots + P_N(1 + (n-1)\lambda)\} \longrightarrow$

$$P_N^\lambda = \sum_{\nu = k\lambda} b_\nu^N e^{ik\lambda \cdot} \quad .$$

This shows that the limit exists and P_N^λ approaches this limit as $N \to \infty$. The rest of the proposition follows immediately.

CHAPTER II

KEPLER AND AFTER

§1. <u>Kepler</u>. It would be difficult to overemphasize the importance of Kepler's work and its impact on the development of science . Copernicus (and apparently the greek astronomer Aristarchus much before him) had introduced the idea of describing the motion of the planets in circles and epicycles about the sun. However , the Copernicus theory in its detailed execution was a complicated (if not more) than the Ptolemaic theory. Furthermore, even with all its complications, it did not adequately explain the motion of the planets to within the experimental accuracy of the time. Finally, it was still a purely geometric construction, which while it did offer a somewhat unified approach to the motion of the planets , still treated each planet separately. Much as we indicated in the last chapter on the Ptolemaic theory of the moon, the planets

94.

were not regarded as "real" objects in space interacting
with each other but as abstract objects whose motions on
the celestial sphere were described by the geometric
constructions.

Kepler's goal was to obtain a description of the
planetary motions on the basis of what he called "causation".
Thus the title of his magnum opus is

"A new astronomy based on causation or a physics
of the sky derived from the investigations of the motions
of the star Mars founded on the observations of the noble
Tycho Brahe".

Kepler's principal goal was to explain the relationship between the existence of five planets (and their motions)
and the five regular solids. It is customary to sneer at
Kepler for this and maintain that he was unaware of the
important parts in his own work. Perhaps two comments
are in order concerning this attitude. One is that it is
instructive to compare this with the current attempts to
"explain" the zoology of elementary particles in terms of
irreducible representations of Lie groups. A second point
is that perhaps this approach of Kepler, to understand the
motions of the planets in relation to each other is the most
important of Kepler's contribution. His idea was that the

motion of each planet was describable in terms of general physical laws applicable to all of them. Thus one might state "Kepler's zeroth law" as asserting that there is some flow on an abstract manifold projecting onto three dimensional space and the motion of the planets are curves which are projections of specific trajectories of the flow. In other words one could conceive of imaginary planets which would then have laws of motion given by Kepler's three laws. Thus the whole notion of describing the motion by a system of differential equations is implicit in Kepler's assumptions. It is only in this context that it makes sense to ask why particular trajectories are observed and not others (a question which remains unanswered to this day).

Kepler set himself the task of describing the motion of the planets, especially Mars and Earth about the sun. Now the actual motion of a planet about the sun is not observed. All that is observed is the direction of the planet and of the sun.

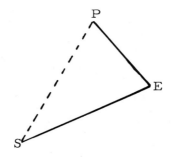

Thus the direction of ES and EP are observed. However, the direction of SP can only be deduced once the ratio PE/SE is known and this cannot be observed directly. In fact the determination of the orbit of SP is indeterminate unless an auxiliary hypothesis is made. Kepler's auxiliary hypothesis was that the planets move on periodic (i. e., closed) orbits around the sun, and that the orbit closes during one revolution about the sun.

Kepler proceeds as follows: first determine the period of the planet. (How this is done will be described in a moment.) Now measure the geocentric position of the planet when it is in opposition to the sun. AE this instant, the heliocentric longitude of the planet is the opposite of the geocentric longitude of the sun, i. e., S, E and P determine a fixed plane orthogonal to the ecliptic. By the assumption of periodicity, if P lies in a given plane at time t_0 it lies in the same plane at time $t_0 + n\tau$ where τ is the period of the planet. By observing many oppositions one determines the longitude of P about the sun at sufficiently close dates to be able to interpolate. One thus obtains the heliocentric longitude of P as a function of time. Now one takes an observation of P when it is not in opposition. This then clearly determines the heliocentric latitude and the ratio EB/SP . Making many observations at the same date of the year (so that ES is a constant) one determines

the shape of the curve SP . Making many observations at

times $t_0 + n\tau$ so that SP is constant, one determines the

curve SE (i. e. , the earth's orbit) . (Actually Kepler

followed a slightly different procedure of successive

approximations which is a little complicated to describe

here.)

    The period of a planet is determined in two in-

dependent ways: in the first procedure one observes several

oppositions. By observing how many times the earth has

overtaken the planet one can compute the average planetary

velocity provided the oppositions are spaced far enough

apart. A second procedure is to determine the period of

return of the planet to the ascending node (one of the points

of intersection of the orbit with the ecliptic) .

    In this way after a multitude of tedious computations

with arithmetical errors cancelling each other out, the

orbit of Mars is drawn .

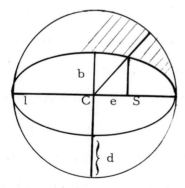

The eccentricity in this
diagram is greatly
exaggerated.

It still took Kepler quite a while to recognize the orbit as an ellipse. The most conspicuous parameters of the orbit (taking the semidiameter as one) are the eccentricity, e, and the maximal diameter of the "sickle" $d = 1 - b$. His figures gave $e = .0926$ and $d = .00429$. He observes that $.00429$ is one half $(.0926)^2 = .00857$, i.e., $d = \frac{1}{2} e^2$. He writes "it was as if I woke from sleep and saw a new light". It takes a little while longer to recognize $\frac{1}{2} e^2$ as $\sim 1 - (1 - e^2)^{1/2}$ and to identify the curve as an ellipse. In any event, Kepler arrived at his three famous laws:

1. The planetary orbits are in planes and are ellipses with the sun as focus.

2. The area swept out by the radius from the sun to the plane depends linearly on the time,

and 3. The square of the periods are proportional to the cubes of the major axes.

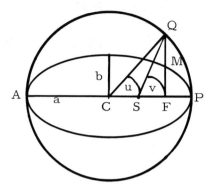

§ 2.   <u>Elliptic Motion</u>. Let us fix notation that we will be using later on.   C   denotes the center of the ellipse whose major semiaxis is   a   and whose eccentricity is   e. Thus S (the sun) is located at one focus of the ellipse and

$$CS \;=\; ae$$

$$CA = CQ \;=\; a$$

and

$$\frac{MF}{QF} \;=\; \frac{b}{a} \;=\; (1 - e^2)^{1/2}$$

The angle   u = PCQ   is called the eccentric anomaly and the angle   PSM = v   is called the true anomaly .

The 2nd law (of areas) says that area   PSM = kt for some constant k .   Now

$$\text{area PSM} = \frac{b}{a} \text{ area PSQ}$$

$$= \frac{b}{a}(\text{area PCQ} - \text{area SCQ})$$

$$= \frac{b}{a}\left(\frac{a^2}{2}u - \frac{a^2 e \sin u}{2}\right)$$

and so we obtain

(2.1) $u - e \sin u = \mu t$ (Kepler's equation)

with $\mu = \dfrac{2k}{ab}$.

The study of methods of solution of the transcendental Kepler's equation provided the impetus for much of the development of the theory of functions of a complex variable. In fact, some of Cauchy's famous memoirs on power series were inspired by the problem of justifying series techniques of solving this equation.

We will call

$$\ell = u - e \sin u$$

the mean anomaly so that $\ell = \mu t + \alpha$ where μ is the mean motion of the planet.

We continue to derive elementary geometric facts concerning elliptical motion, primarily to establish notation to be used later on. Introducing polar and rectangular

coordinates with the sun as origin we have

$$x = r \cos v$$
$$y = r \sin v \quad .$$

For (x, y) on the ellipse we have

$$x = a(\cos u - e)$$
$$y = \frac{b}{a} \cdot a \sin u = b \sin u \quad .$$

Thus

$$r^2 = a^2[\cos^2 u + e^2 - 2e \cos u + (1 - e^2) \sin u]$$
$$= a^2[1 + e^2 \cos^2 u - 2e \cos u]$$

or

$$(2.2) \qquad\qquad r = a(1 - e \cos u)$$

is the equation of the ellipse where r is given as a

function of the eccentric anomaly. Now

$$x = r \cos v = a(\cos u - e)$$

so

$$r(1 + e \cos v) = a(1 - e \cos u) + ae(\cos u - e)$$

thus

$$r(1 + e \cos v) = a(1 - e^2)$$

or

(2.3) $$r = a \frac{1 - e^2}{1 + e \cos v}$$

is the equation of the ellipse in polar coordinates.

We now compute the velocity as given by Kepler's second law. We have

$$dx^2 = a^2 \sin^2 u \, du^2$$

$$dy^2 = b^2 \cos^2 u \, du^2 = (1 - e^2)a^2 \cos^2 u \, du^2$$

so that

$$dx^2 + dy^2 = a^2(1 - e^2 \cos^2 u) \, du^2 \, .$$

Subtracting (2.2) from 2a we get

$$2a - r = a(1 + e \cos u)$$

and multiplying by (2.2) yields

$$r(2a - r) = a^2(1 - e^2 \cos^2 u)$$

so

$$dx^2 + dy^2 = r(2a - r) du^2 \, .$$

By Kepler's equation

$$\mu \, dt = (1 - e \cos u) du = \frac{r}{a} \, du$$

or

$$du = \frac{\mu a}{r} dt \quad .$$

Thus

(2.4) $\qquad dx^2 + dy^2 = \mu^2 a^3 \left[\frac{2}{r} - \frac{1}{a} \right] dt^2 \quad .$

§3. <u>Newtonian mechanics</u>. There is no point repeating here the basic facts concerning Newton's laws and their importance for the development of mathematics and science. They are treated in any elementary physics book. In this section we briefly point out their connection with some of the formulas of the last section. Let m be the mass of a planet. According to the principle of conservation of energy we must have $\frac{d}{dt}(U + T) = 0$ where U is potential and T is kinetic energy. It follows from Kepler's laws (see any elementary physics text) that $U = U(r)$. By (2.4) we see that

(3.1) $\qquad T = \frac{1}{2} mv^2 = \frac{m\mu^2 a^3}{2} \left[\frac{2}{r} - \frac{1}{a} \right]$

so that, up to a constant

$$U = - \frac{m\mu^2 a^3}{r} \quad .$$

By symmetry (the "equality of action and reaction")
$U = - \dfrac{KmM}{r}$ where K is some universal constant and
M is the mass of the sun. Thus

(3. 2) $$a^3 \mu^2 = KM .$$

Equation (3.1) allows the (approximate) determination of the ratio of the mass of any large planet having a
satellite to the mass of the sun . In fact, let a' and μ'
be the semi-major axis and mean motion of the satellite
about the planet, and let M' be the mass of the planet.
Then

(3. 3) $$\dfrac{M'}{M} = \left(\dfrac{a'}{a}\right)^3 \left(\dfrac{\mu'}{\mu}\right)^2 .$$

This of course, neglects the effect of the sun on the
motion of the satellite.

§4. The three body problem. Let us now consider
the equations of motion of two planets, P and P_1 and
the sun, moving under Newton's laws and ignoring all
other effects.

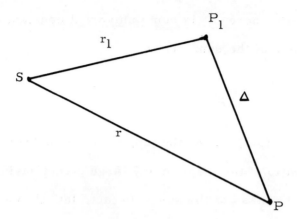

Let (ξ_S, η_S, ξ_S) be the coordinates of the sun, (ξ_P, η_P, ξ_P) be the coordinates of P and $(\xi_{P_1}, \eta_{P_1}, \xi_{P_1})$ be the coordinates of P_1. The force on the sun in the ξ direction is (setting the gravitutional constant $K = 1$),

$$\frac{Mm(\xi_P - \xi_S)}{r^3} + \frac{Mm_1(\xi_{P_1} - \xi_S)}{r_1^3}$$

so the sun's acceleration is

$$\frac{m(\xi_P - \xi_S)}{r^3} + \frac{m_1(\xi_{P_1} - \xi_S)}{r_1^3}$$

while the acceleration of P is

$$\frac{M(\xi_P - \xi_S)}{r^3} + \frac{m_1(\xi_P - \xi_{P_1})}{\Delta^3} \ .$$

Choosing S as the origin of coordinates, i. e.,
setting $x = x_P = \xi_P - \xi_S$ and $x_1 = x_{P_1} = \xi_{P_1} - \xi_S$ we see
that

$$(4.1) \qquad \ddot{x} = - \frac{(M + m)x}{r^3} - \frac{m_1 x_1}{r_1^3} - \frac{m_1(x - x_1)}{\Delta^3} \ .$$

By symmetry

$$(4.2) \qquad \ddot{x}_1 = - \frac{(M + m_1)x_1}{r_1^3} - \frac{mx}{r^3} - \frac{m_1(x_1 - x)}{\Delta^3} \ .$$

If we set

$$(4.3) \qquad V = \frac{m_1(m + M)}{r} + \frac{mm_1}{\Delta} - \frac{mm_1(x_1 x + y_1 y + z_1 z)}{r_1^3} \ ,$$

we can write

$$m \frac{d^2 x}{dt^2} = \frac{\partial V}{\partial x} \qquad m \frac{d^2 y}{dt^2} = \frac{\partial V}{\partial y} \qquad m \frac{d^2 z}{dt^2} = \frac{\partial V}{\partial z} \ .$$

Let us set $p_x = m\dot{x}$, $p_y = m\dot{y}$, $p_z = m\dot{z}$,

$$T = \frac{1}{2m} (p_x^2 + p_y^2 + p_z^2)$$

and $H = T - V$ we can write these equations in the form

$$(4.4) \qquad \frac{dx}{dt} = \frac{\partial H}{\partial p_x} \qquad \frac{dp_x}{dt} = - \frac{\partial H}{\partial x}$$

$$\frac{dy}{dt} = \frac{\partial H}{\partial p_y} \qquad \frac{dp_y}{dt} = - \frac{\partial H}{\partial y}$$

$$\frac{dz}{dt} = \frac{\partial H}{\partial p_x} \qquad \frac{dp_z}{dt} = - \frac{\partial H}{\partial z} \qquad .$$

Such equations are said to be in canonical form. We can, of course, find similar equations for P_1 .

§5. <u>The method of Lagrange</u>. Notice that in (4.1) if we set $m_1 = 0$, we obtain the equations of motion of a particle attracted by a mass $M + m$ at the origin. According to Kepler's laws, this motion is elliptical. A specific ellipse would be specified by five parameters:

i = inclination of the plane of the ellipse (relative to a fixed plane, say the ecliptic).

θ = longitude of the line of nodes in the ecliptic

g = angle of the perihelion in the orbital plane to the line of nodes

a = major semi-axis

e = eccentricity .

To specify the actual position of the planet we must give an angular coordinate on the ellipse and we take it to be the mean anomaly

$$\ell = u - e \sin u \quad .$$

Thus if m_1 were zero then i, θ, g, a, e would be constants and ℓ a linear function of time. Lagrange's idea was that if m_1 is small then the motion (for any finite period of time) will not deviate much from Kepler motion (at least if r_1 and δ_1 are not too small). Now x, y, z, p_2, p_y, p_z are expressible as functions of i, θ, g, a, e, ℓ granting Kepler's laws (so that the velocity is determined at any point of the ellipse.) Therefore if we make a change of variables, the differential equations (4.4) are transformed into equations involving the six elliptical elements, and the derivatives for all terms except ℓ should be small. If we now assume that m is also small then, by (4.2), P_1 will not deviate much from an elliptical motion. Therefore, if we substitute an approximate elliptic curve for P_1 into the equations for the elements of P we will make an error of second order in the sizes of the masses. In carrying out this program, a number of interesting features arise.

To simplify the calculations we shall replace the coordinates i, θ, g, a, e, ℓ by another system (Delaunay elements) in terms of which the equations take a singular form. In this presentation we follow Poincaré [P7]. Our variables are as follows: we keep θ, ℓ and g, and set

(5.1)
$$L = m(M + m)^{1/2} a^{1/2}$$

(5.2)
$$G = L(1 - e^2)^{1/2}$$

(5.3)
$$\Theta = G \cos i \quad .$$

Notice that by (3.1) for a given Kepler motion we have ((3.2) with M replaced by M + m and K = 1)

$$\frac{1}{2} m v^2 = \frac{m(M + m)}{2}\left(\frac{2}{r} - \frac{1}{a}\right)$$

so that

(5.4)
$$\frac{1}{2} m v^2 - \frac{m(M + m)}{r} = - \frac{m(M + m)}{2a} = - \frac{m^3(M + m)^2}{2L^2} .$$

Also

$$G = m(M + m)^{1/2} a^{1/2}(1 - e^2)^{1/2}$$

and by (3.2) (where K = 1)

$$\mu = (M + m)^{1/2} a^{-3/2}$$

while $\mu = \dfrac{2k}{ab}$ by the argument leading to (2.1). Thus

$$k = \frac{ab}{2}\mu = \frac{(1 - e^2)^{1/2} a^2 \mu}{2} = \frac{(1 - e^2)^{1/2} a^{1/2}(M + m)^{1/2}}{2}.$$

Thus

(5.5) $$G = \frac{1}{2}mk$$

where k is the rate at which area is swept out, i.e., the angular velocity.

The purpose of introducing the variables θ, ℓ, g, Θ, L, G is the following pair of facts. Both of these facts can be verified by direct substitution, but will be easy consequences of the Hamiltonian formalism introduced in Chapter IV and so we ask the reader to take them on faith.

(i) If we express x, y, z, p_x, p_y, p_z in terms of $\ell, g, \theta, L, G, \Theta$ it turns out that

$$x dp_x + y dp_y + z dp_z - L d\ell - G dg - \Theta d\theta$$

is an exact differential.

(ii) Any change of variables satisfying (i) preserves the form of equations (4.4). That is, the equations in the new coordinates takes the form

$$\frac{d\ell}{dt} = \frac{\partial E}{\partial \ell} \qquad\qquad \frac{dL}{dt} = -\frac{\partial E}{\partial \ell}$$

(5.6) $$\frac{dg}{dt} = \frac{\partial E}{\partial G} \qquad\qquad \frac{dG}{dt} = -\frac{\partial E}{\partial g}$$

$$\frac{d\theta}{dt} = \frac{\partial E}{\partial \Theta} \qquad\qquad \frac{d\Theta}{dt} = -\frac{\partial E}{\partial \theta} \quad ,$$

where $E(\theta, \ell, g, \Theta, L, G) = H(x, y, z, p_x, p_y, p_z)$ when the $x, \cdots, p_z$ are expressed as functions of $\theta, \cdots, G$.

Thus, if $m_1 = 0$ then (5.4) shows that

$$E(\theta, \ell, g, \Theta, L, G) = -\frac{m^2(M+m)^2}{2L^2} \quad .$$

In this situation (5.6) yields the desired fact that all the variables must be constant except ℓ which varies linearly with time.

If $m_1 \neq 0$ we can still write

(5.7) $$E = -\frac{m^3(M+m)^2}{2L^2} - \frac{m\,m_1}{\Delta} + \frac{m\,m_1}{r_1^3}(x\,x_1 + y\,y_1 + z\,z_1)$$

where x, x_1 etc., Δ and r_1 are to be expressed in terms of the new coordinates. In fact, we shall introduce coordinates $\theta_1, \ell_1, g_1, \Theta_1, L_1, G_1$ for the planet P_1 as well.

We now observe that the coordinates ℓ and ℓ_1 enter as angular variables. Thus E is a periodic function of ℓ and ℓ_1 . We can therefore write

$$E = - \frac{m^3(M+m)^2}{2L^2} + \sum A_{p, p_1} e^{i(p\, \ell + p_1 \ell_1)}$$

where p, p_1 range over all pairs of integers. Here the coefficients A_{p, p_1} are functions of the remaining 10 variables, and they all are of the order of magnitude $m\,m_1$.

Now let us examine any equation of (5.6) except the two on the first line. For example in

$$\frac{dg}{dt} = \frac{\partial E}{\partial G}$$

only terms involving the derivatives of the A's enter. Therefore, if in the right hand side, we replace the actual solution curves by the curves

$$\ell = \mu t + \alpha$$

$$\ell_1 = \mu_1 t + \alpha_1$$

all the other variables constant, we will (for any fixed interval of time) have made an error of the order of magnitude m or m_1 so the total effect on the motion will change by a term involving $m^2 m_1$ or $m\,m_1^2$. Thus, up to this order of magnitude, we will have

$$\frac{dg}{dt} = \sum \frac{\partial A_{p, p_1}}{\partial G} e^{i[(p\mu + p_1\mu_1)t + p\alpha + p_1\alpha_1]} \, .$$

Thus, assuming $p\mu + p_1\mu_1 \neq 0$ for $(p, p_1) \neq (0, 0)$ we can write

$$g = C + \frac{\partial A_{0, 0}}{\partial G} t + \sum_{(p, p_1) \neq (0, 0)} \frac{1}{p\mu + p_1\mu_1} \frac{\partial A_{p, p_2}}{\partial G} e^{i[p\alpha + p_1\alpha_1]}$$

The term $\dfrac{\partial A_{0, 0}}{\partial G} t$ is called the secular term, the others are called periodic terms.

Notice that in the equation for L we have

$$\frac{dL}{dt} = -\sum p A_{p, p_1} e^{i[(p\mu + p_1\mu_1)t + p_1\alpha_1 + p_2\alpha_2]}$$

so no secular term appears. This is Lagrange's famous "proof" of the stability of the solar system: the L is proportional to the major semi-axis and we see that this has no secular variation.

Another point that will be central to our discussion in Chapter III is the effect of the denominators $p\mu + p_1\mu_1$ occurring in the expressions for the elements. Since E is an analytic function the Fourier coefficients A_{p, p_1} and

their partial derivatives decrease rapidly. However, it is
conceivable that $p\mu + p_1\mu_1$ may be very small so that the
term $\dfrac{1}{p\mu + p_1\mu_1} A_{pp_1}$ is not negligible . This happens, for
example with Jupiter and Saturn. The mean motion of
Jupiter is about $\mu = 30°.35$ per year while that of Saturn
is about $\mu_1 = 12°.23$ per year. Thus $2\mu - 5\mu_1$ is very
small although 2 and 5 are not large numbers. Thus
$2\mu - 5\mu_1 = .41$ approximately and therefore there is a
substantial periodic disturbance in the orbit of Jupiter with
period slightly under 900 years. This discovery was one
of the triumphs of this approach to the three body problem.
Until the time of Lagrange a deviation in Jupiter's orbit
from Kepler motion was observed but not understood.
Lagrange established that the deviation is indeed periodic
but with long period and due, essentially to the reason
described above. This general phenomenon is known as the
effect of small divisors. It has long been the source of
convergence difficulties in celestial mechanics. Only with
the recent work of Siegel, Kolmogorov, Arnold, and Moser,
to be described in Chapter III, have these difficulties been
surmounted.

In order to emphasize how the arithmetic enters
into the distortion of the orbit we reproduce an elementary
intuitive explanation as can be found, for example, in the

116.

Encyclopedia Brittanica (11th edition, vol. 11, p. 8054).

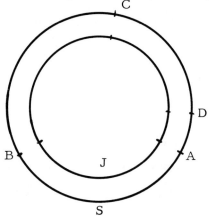

Let the inner circle represent the orbit of Jupiter
and the outer circle the orbit of Saturn. Suppose that the
planets are in conjunction at some position A .

They will next be in conjunction after 20 years when
Jupiter will have moved about 600° and Saturn 240° and
this will occur at point B . After another 20 years they
will arrive again at conjunction at C and then at D . If
the mean motions were exactly 30° per year and 12° per
year then D would coincide with A . Then periodically
there would be an attraction of Saturn on Jupiter which
would not be balanced out and would thus disturb the shape
of the orbits. Actually, D ≠ A and after a long time the
conjunctions become dense in the circle and balance things
out .

One final comment about Lagrange's method is that it is only valid for a fixed finite interval of time, whose length depends on the ratio of the masses. In particular, the "secular terms" involving known expressions in time only refer to the given approximation. They do not imply that the corresponding elements are unbounded, for instance. This is because when they exceed a certain bound, the approximation assumptions (of assuming constant values for the elements in E) are no longer justibiable. In fact, the main theorem of Arnold [A6] asserts that for "most" positions, the motion is almost periodic.

§6. Hill's lunar theory and the restricted three body problem. Let us reconsider the situation similar to the earth-moon-sun system. Thus now S becomes the earth, P becomes the moon and P_1 the sun. We still have equation (4.1)

$$\ddot{x} = - \frac{\left(M_{(earth)} + m_{(moon)}\right)x}{r^3} - \frac{m_{1(sun)}x_1}{r_1^3}$$
$$- \frac{m_{1(sun)}(x - x_1)}{\Delta^3}$$

with corresponding equations for y and z . Here r is the distance from the moon to the earth, r_1 the distance

from the sun to the earth and Δ the distance from the moon
to the sun . We also obtain the corresponding equation (4.2).

Because $r \ll r_1$, and the physical facts that the
barycenter of the earth-moon system lies inside the earth
and that of the earth-sun lies inside the sun, it follows that
the perturbation of the sun's orbit is very small - so that
we may consider the sun's motion elliptical. Sun's effect
on the moon's orbit however is not so negligible, and this
is the cause of difficulty in the lunar theory. Let φ denote
the angle

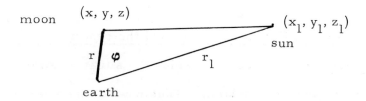

between r and r_1 so that

$$rr_1 \cos \varphi = xx_1 + yy_1 + zz_1 .$$

Thus

$$\Delta^2 = (x - x_1)^2 + (y - y_1)^2 + (z - z_1)^2 = r^2 + r_1^2 - 2rr_1 \cos \varphi .$$

Let us write $r = \alpha r_1$ so α is a function of time which is
small ($\sim 1/400$) . We can thus expand

$$\frac{1}{\Delta} = \frac{1}{r_1}(1 - 2\alpha \cos \varphi + \alpha^2)^{-1/2}$$

$$= \frac{1}{r_1} \sum \mathcal{L}_k(\varphi)\alpha^k$$

$$= \frac{1}{r_1} + \frac{\alpha \cos \varphi}{r_1} + \cdots \frac{\alpha^2}{r_1}\left[-\frac{1}{2} + \frac{3}{2}\cos^2\varphi\right]$$

$$+ \cdots .$$

Now let us use the form $\ddot{x} = \frac{1}{m}\frac{2V}{2x}$ where, by (4.3),

$$\frac{1}{m}V = \frac{M + m}{r} + \frac{m_1}{\Delta} - \frac{m_1}{r_1^3} r r_1 \cos \varphi .$$

Writing $\dfrac{m_1 r}{r_1^2} \cos \varphi$ as $\dfrac{m_1 \alpha}{r_1} \cos \varphi$ we see that

$$\frac{m_1}{\Delta} - \frac{m_1}{r_1^3}rr_1 \cos \varphi = \frac{m_1}{r_1} + \frac{m}{r_1}\alpha^2\left[-\frac{1}{2} + \frac{3}{2}\cos^2\varphi\right] + \cdots .$$

Now $\dfrac{m_1}{r_1}$ doesn't depend on (x, y, z) and so can be omitted from V. Thus the major perturbing term is

$$\frac{m_1}{r_1}\alpha^2\left[-\frac{1}{2} + \frac{3}{2}\cos^2\varphi\right] .$$

Comparing it to the principal term $\dfrac{m_1 + M}{r}$ we see that the quotient is proportional to

$$\frac{m_1}{M+m} \alpha^3 = \frac{m_1}{M+m} \frac{a^3}{a_1^3} \left(\frac{r}{a}\right)^3 \left(\frac{a_1}{r_1}\right)^3 ,$$

(where a is the major sum axis if the moon and a_1 of the sun). By (3.2) (with $K = 1$) this is approximately

$$\frac{m_1 a^3}{(M+m)a_1^3} = \frac{\mu_1^2}{\mu^2}$$

where μ_1 is the (approximate) mean motion of the sun (= 1 per year) while μ is that of the moon (= between 12 and 13 per year). Thus the major term in the perturbation is of order of magnitude $\frac{1}{12} \cdot \frac{1}{13}$.

Hill, in his fundamental and epoch making paper handles the problem as follows: Let us assume, to first approximation, that we can ignore the solar eccentricity $(e_1 = .017)$ so $\frac{r_1}{a_1} = 1$ and the ratio $\alpha \, (= \sim .0025)$. (These effects can be reintroduced later as perturbations.) Then the perturbing function is

$$\frac{m_1}{r_1} \alpha^2 \left[-\frac{1}{2} + \frac{3}{2}\cos^2\varphi \right] \cong \frac{m_1 r^2}{a_1^3} \left[-\frac{1}{2} + \frac{3}{2}\cos^2\varphi \right]$$

$$= \mu_1^2 r^2 \left[-\frac{1}{2} + \frac{3}{2}\cos^2\varphi \right] .$$

Let us now introduce coordinates X, Y, Z where $Z = z$ and X, Y rotate in the $x\,y$ plane (which is the ecliptic) with angular velocity μ_1 so that the positive X axis is always directed toward the (mean) sun .

Let us rewrite the equations of motion in these new coordinates. Evidently

$$\begin{bmatrix} X \\ Y \end{bmatrix} = A(t) \begin{bmatrix} x \\ y \end{bmatrix} \quad \text{where} \quad A(t) = \begin{bmatrix} \cos \mu_1 t & \sin \mu_1 t \\ -\sin \mu_1 t & \cos \mu_1 t \end{bmatrix} .$$

Now,

$$\dot{A}(t) = \mu_1 \begin{bmatrix} -\sin \mu_1 t & +\cos \mu_1 t \\ -\cos \mu_1 t & -\sin \mu_1 t \end{bmatrix} \text{ and } \ddot{A}(t) = -\mu_1^2 A(t) .$$

Hence

$$\begin{bmatrix} \dot{X} \\ \dot{Y} \end{bmatrix} = \dot{A}(t) \begin{bmatrix} x \\ y \end{bmatrix} + A(t) \begin{bmatrix} \dot{x} \\ \dot{y} \end{bmatrix} = \mu_1 \begin{bmatrix} Y \\ -X \end{bmatrix} + A(t) \begin{bmatrix} \dot{x} \\ \dot{y} \end{bmatrix}$$

and so

$$\begin{bmatrix} \ddot{X} \\ \ddot{Y} \end{bmatrix} = \ddot{A}(t) \begin{bmatrix} x \\ y \end{bmatrix} + 2 \dot{A}(t) \begin{bmatrix} \dot{x} \\ \dot{y} \end{bmatrix} + A(t) \begin{bmatrix} \ddot{x} \\ \ddot{y} \end{bmatrix}$$

$$= -\mu_1^2 \begin{bmatrix} X \\ Y \end{bmatrix} + 2 \dot{A}(t) \, A(t)^{-1} \begin{bmatrix} \dot{X} - \mu_1 Y \\ \dot{Y} + \mu_1 X \end{bmatrix}$$

$$+ A(t) \begin{bmatrix} \dfrac{\partial V}{\partial x} \\[2mm] \dfrac{\partial V}{\partial y} \end{bmatrix} .$$

It is easily seen that

$$\dot{A}(t)\, A(t)^{-1} \;=\; \begin{bmatrix} 0 & 1 \\ -1 & 0 \end{bmatrix} ,$$

and the last term equals

$$\begin{bmatrix} \dfrac{\partial V}{\partial X} \\[2mm] \dfrac{\partial V}{\partial Y} \end{bmatrix} .$$

So finally we have

$$\frac{d^2 X}{dt^2} - 2\mu_1 \frac{dY}{dt} - \mu_1^2 X = -\frac{(M+m)X}{r^3} + \frac{\partial R}{\partial X}$$

(6.1)
$$\frac{d^2 Y}{dt^2} + 2\mu_1 \frac{dX}{dt} - \mu_1^2 Y = -\frac{(M+m)Y}{r^3} + \frac{\partial R}{\partial Y}$$

$$\frac{d^2 Z}{dt^2} = -\frac{(M+m)Z}{r^3} + \frac{\partial R}{\partial Z}$$

where

$$R = \mu_1^2\, r^2 \left[-\frac{1}{2} + \frac{3}{2} \cos^2 \varphi \right] .$$

Let us make the further assumption that the lunar inclination is small. We will therefore, to first approximation, set $r \cos \varphi = X$ and the first two equations become

$$\frac{d^2X}{dt^2} - 2\mu_1 \frac{dY}{dt} - 3\mu_1^2 X + \frac{(M+m)X}{r^3} = 0$$

$$\frac{d^2Y}{dt^2} + 2\mu_1 \frac{dX}{dt} + \frac{(M+m)Y}{r^3} = 0 \ .$$

In order to compare the magnitudes of the various terms it is standard practice to make a change of scale $\tau = (\mu - \mu_1)t$ so denote by $\dot{X}$ and $\ddot{X}$ derivatives with respect to τ we get

$$\ddot{X} - 2\nu\dot{Y} - 3\nu^2 X + \frac{KX}{r^3} = 0$$

(6.2)

$$\ddot{Y} + 2\nu\dot{X} + \frac{KY}{r^3} = 0$$

where

$$\nu = \frac{\mu_1}{\mu - \mu_1} \qquad (\sim \tfrac{1}{12} \text{ in the case of the moon})$$

and

$$K = \frac{(M+m)}{(\mu - \mu_1)^2} \ .$$

Hill's procedure was to construct an explicit periodic solution of (6.2) which is close to the actual solution of (6.2) corresponding to our approximation to the moon's orbit. This periodic solution is called the "variational curve". Since the desired solution is to be close to the variational curve, it can (to a high degree of approximation) be found by solving the "equations of variation" of (6.2) along this curve. (Cf.[L3] pp. 513-515 for the definition and elementary properties of the equations of variation.)

Notice that ν is small. If we could ignore ν^2 then an explicit solution would be the circle

$$X = a \cos \lambda \tau \qquad Y = a \sin \lambda \tau$$

if

$$\frac{K}{a^3} = \lambda^2 + 2\lambda\nu \quad .$$

It is to be expected that the presence of the term $-3\nu^2 X$ would have the effect of distorting this circle into an oval shaped curve. In fact observe that if $(X(\tau), Y(\tau))$ is a solution of (6.2), so are $(X(-\tau), -Y(-\tau))$ and $(-X(-\tau), Y(-\tau))$. The first of these facts implies, by the uniqueness of differential equations, that if $(X(\tau), Y(\tau))$ is a solution such that $Y(0) = 0$, $\dot{X}(0) = 0$ (so the curve crosses the x-axis at right angles) then $X(-\tau) = X(\tau)$, $Y(-\tau) = -Y(\tau)$. Thus the curve is symmetrical with respect to the x-axis. Similarly, a curve crossing the Y axis at right angles will be symmetrical with respect to the Y axis.

Now consider solution curves of (6.2) which satisfy $Y(0) = \dot{X}(0)$, but with varying values of $\dot{Y}(0)$. Suppose that some of these curves cross the Y axis at angles $< \frac{\pi}{2}$

and others cross at angles $> \frac{\pi}{2}$. Then at some inter-
mediate value of $\dot{Y}(0)$ it will cross the Y axis at right
angles and the curve will be symmetrical with respect to
both axes, hence a closed curve, in fact an oval. (One
could establish that by suitably varying $\dot{Y}(0)$ the angle of
crossing ranges over an interval containing $\frac{\pi}{2}$ by first
establishing this result for the equations with $3\nu^2 X$
missing (by applying the equations of variation) and then,
regarding $3\nu^2$ as a small parameter, conclude the same
for (6.2)). We can, in fact, conclude the existence of
such periodic curves by the methods we shall present in
Chapter IV for the restricted three body problem.

Guided by the fact that the curve is an oval, Hill
proceeded to determine the periodic solution by setting

(6.3)
$$X(\tau) = \Sigma \; A_n \cos (2n + 1)\omega\tau$$
$$Y(\tau) = \Sigma \; B_n \sin (2n + 1)\omega\tau$$

and determining the coefficients A_n and B_n . (Here no
sine terms occur in $X(\tau)$ or cos terms in $Y(\tau)$ since
$X(-\tau) = X(\tau)$ and $Y(-\tau) = - Y(\tau)$. Similarly only odd
integers enter because of the symmetry with respect to
the Y-axis. Here ω is determined to be the empirically
known mean synodic period of the moon (the mean month).

By substituting (6.3) into (6.2) he obtained an infinite
system of non-linear equations in an infinite number of
unknowns, which he proceeded to solve. (For a justifica-
tion of this method see Wintner [W 10]. An alternative
construction of this "variational curve" can be found in
Siegel[S 3] .

As we already remarked, the actual desired solution
of (6.2) is obtained from the given one by studying the
equations of variation. This will then be a (second order)
linear differential equation with periodic coefficients. We
must also take into account the third equation of (6.1) . It
turns out, (to first order in the lunar inclination) that these
equations have similar forms; we examine this last equation.

Suppose we ignore the square of the inclination. This
is equivalent to still having $X = r \cos \varphi$ so that the first
two equations of (6.1) remain independent of the third and
we will therefore assume that X and Y are the solutions
of (6.2) given by the variational curve. Thus X and Y
are known periodic functions of τ and the third equation
can be written as

$$(6.4) \qquad \qquad \ddot{Z} + \Theta Z = 0$$

where $\Theta = \nu^2 + \dfrac{K^2}{r^3}$. Now if we multiply the first equation of (6.2) by $\dot{X}$ and the second by $\dot{Y}$ and add we get

$$\frac{d}{d\tau}\ \frac{1}{2}(\dot{X}^2 + \dot{Y}^2) - \frac{3}{2}\nu^2(X^2) - \left(\frac{K}{r}\right) = 0$$

or $\frac{1}{2}(\dot{X}^2 + \dot{Y}^2) - \frac{3}{2}\nu^2 X^2 - \frac{K}{r} = C$ (= constant) which is known as Jacobi's integral. Thus

$$\frac{K}{r} = \frac{1}{2}\left(\dot{X}^2 + \dot{Y}^2\right) - \frac{3}{2}\nu^2 X^2 - C\ .$$

Since X and Y are trigonometric series in $\cos(2n+1)\omega\tau$, $\sin(2n+1)\omega\tau$ it follows from the Jacobi integral that Θ is a trigonometric series with only even coefficients. Since $\tau \to -\tau$ leaves the right side invariant, Θ involves only cosines. We can thus write

$$\Theta = \Theta_0 + 2\Theta_1 \cos 2\omega\tau + 2\Theta_2 \cos 4\omega\tau + \cdots \ .$$

Hill's procedure for solving (6.4) was the following: Write the equation as

$$\frac{d^2 z}{ds^2} + z \sum_{-\infty}^{\infty} q_\alpha e^{2i\alpha s} = 0$$

(where we have made an appropriate change of time scale again to simplify the equations). Look for a solution in

128.

the form

$$Z = \sum_j b_j e^{i(\mu+2j)s} \quad .$$

Then (6.4) becomes

$$-\sum(\mu + 2j)^2 b_j \, e^{i(\mu+2j)s} + \sum b_j e^{i(\mu+2j)s} \sum q_\alpha \, e^{2i\alpha s} \quad ,$$

or collecting and comparing coefficients,

$$(\mu + 2j)^2 b_j = \sum b_i q_{j-i} \quad .$$

He was thus led to an infinite system of linear equations.
He introduced a so called "infinite determinant" to solve
this equation. This procedure was then justified by
Poincaré, [P 3] and [P 4] .

Thus in addition to giving a masterly and close
approximation to the moon's motion, Hill's work introduced
three key ideas which were taken up by Poincaré and
developed into major areas of modern mathematical thought:
the idea of qualitative geometric study of differential
equations, implicit function methods in infinite dimensional
spaces, and linear analysis in infinite dimensional spaces.

It is interesting to point out that Hill did not hold any
academic position in this country. He was a clerk at the
National Bureau of Standards, working long hours and doing

his own research at night. He went unappreciated by his

colleagues for many years. When Poincaré visited the

United States after the turn of the century, the one person

he went to visit was Hill. Hill was then offered a promotion

but declined on the grounds that increased responsibility

would leave him less time for his own work. Simon

Newcomb[N5] described Hill as "perhaps the greatest

living master in the highest and most difficult field of

astronomy, winning world wide recognition for his country

in the science, and receiving the salary of a department

clerk".

§7.   The general theory of relativity. There exist

many books on general relativity so here we include a very

sketchy discription of what as pronounced predictions are

made by the theory, referring the reader to any of the

standard books for the details(my favorite is the one by

Synge[S14].We recall that according to general relativity the

four dimensional universe processes a metric tensor

$\Sigma_{i,j} \, g_{ij} dx^i dx^j$ of signature $---+$, so that the tangent space

at each point carries a Lorentz metric. The four di-

mensional universe represents all events. Each particle

(during its entire history) gives rise to a curve $x^i(t), i \to t$

$= 1, 2, 3, 4$. It is assumed that all such curves (i.e., curves corresponding to actual particle histories) are "time-like" in the sense that

$$\sum_{i,j} g_{ij} \frac{dx^i}{dt} \frac{dx^j}{dt} \geq 0 \quad .$$

Thus all such curves have a well defined "length" which has the following physical meaning: It is assumed that each particle for which

$$\sum g_{ij} \frac{dx^i}{dt} \frac{dx^j}{dt} > 0$$

has attached to it a fundamental measure of time (determined up to scale) called proper time. (It would be measured, say, by an atomic clock carried by the particle.) Then

$$\int_{t_1}^{t_2} \sqrt{\sum g_{ij} \frac{dx^i}{dt} \frac{dx^j}{dt}} \, dt$$

is the length of time between t_1 and t_2 .

We recall some facts from geometry [S13]: An affine connection is a differential operator from $T(M) \to T \otimes T^*(M)$ such that $D(fY) = fDY + Y \otimes df$. Another way of putting it is to say that an affine connection is a rule which assigns, to each vector field X , a differential operator ∇_X on vector fields so that $Y \rightsquigarrow \nabla_X Y$ is a linear map on vector

$\nabla_X \cdot Y$ is additive in X and Y

with

$$\nabla_{gX} Y = g\nabla_X Y$$

and

$$\nabla_X gY = g\nabla_X Y + (Xg)Y \ .$$

The fundamental assertion is that there is a unique torsion-less connection preserving the Lorenzian metric, i. e., such that

$$(\nabla_X Y, Z) + (Y, \nabla_X Z) = X\{(Y, Z)\}$$

where (,) denotes the pointwise scalar product of two vector fields.

In terms of local coordinates $x^1, \cdots, x^4$

$$\nabla_{\frac{\partial}{\partial x^i}} \frac{\partial}{\partial x^j} = \sum \Gamma_{ij}^k \frac{\partial}{\partial x^k}$$

where

$$(7.1) \qquad \Gamma_{ij}^k = \frac{1}{2} \sum_\ell g^{k\ell} \left(\frac{\partial g_{j\ell}}{\partial x^i} + \frac{\partial g_{\ell i}}{\partial x^j} - \frac{\partial g_{ij}}{\partial x^\ell} \right) \ .$$

The geodesics are given by

$$(7.2) \qquad \frac{d^2 x^k}{dt^2} + \sum \Gamma_{ij}^k \frac{dx^i}{dt} \frac{dx^j}{dt} = 0 \ .$$

Let us introduce normal coordinates about some points in space time so that

$$g_{11} = -1 + \gamma_{11} \, , \ g_{12} = \gamma_{12}, \ \cdots, \ g_{44} = 1 + \gamma_{44}$$

where the γ's are small together with the first derivatives with respect to $x^1, \cdots, x^4$ at the point in question. Let us consider the equation for geodesics where (the initial conditions - hence conditions for a short period are such that)

$$\frac{dx^i}{dt} \qquad i = 1, 2, 3 \qquad \text{are small}$$

and

$$\frac{dx^4}{dt} = 1 + \beta \qquad \text{where } \beta \text{ is small}$$

so that we will, to first approximation ignore any quadratic term in γ, Γ^i_{jk}, β or $\frac{dx^i}{dt}$ ($i = 1, 2, 3$). Then, for $k = 1, 2, 3$, we have, approximately

$$\frac{d^2 x^k}{dt^2} = -\Gamma^k_{44} = -\frac{1}{2} \frac{\partial g_{44}}{\partial x^k} + \frac{\partial g_{4k}}{\partial x^4} \quad .$$

Suppose our coordinate system was constructed as follows: Start with a time-like geodesic and exponentiate its normal bundle. Then introduce the orthogonal trajectories to the hypersurfaces corresponding to each fiber. If x^4 is a (geodesic) parameter on the geodesic and x^1, x^2, x^3

linear coordinates on the fiber at $x^4 = 0$, we get coordinates

with $g_{4k} \equiv 0$.

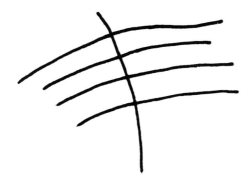

The above equations then become

$$\frac{d^2 x^k}{dt^2} = -\frac{1}{2} \frac{\partial g_{44}}{\partial x^k} \quad .$$

This suggests considering $\frac{1}{2} g_{44}$ as a potential. Thus

Einstein was thus led to the idea that the metric g_{ij}

represents the gravitational field and that a "free" particle

of mass m moves according to

(7.3) $\qquad m \frac{d^2 x^2}{dt^2} + \sum \Gamma^i_{jk} \frac{dx^j}{dt} \frac{dx^k}{dt} = 0$.

In the absence of a gravitational field it is assumed

that the metric is flat. We can thus introduce coordinates

(in fact the ones introduced above) such that the g_{ij} are

constant. This means that the Γ's are zero and a free particle moves with constant velocity. Such a coordinate system is called an inertial coordinate system.

Observe that about any point we can introduce normal coordinates so that at the point in question we can make the $\Gamma^i_{jk} = 0$. Thus we can, for one instant, make the gravitational field "disappear" . This is known as the "principle of equivalence" - at an instant the gravitational force looks the same as the force due to acceleration relative to an inertial frame.

Along a timelike geodesic it is natural to introduce the "proper time", ds, as coordinate. Then solution of (7.3) will have a tangent vector

$$v = \left(\frac{dx^1}{ds} , \ldots , \frac{dx^4}{ds} \right)$$

with length one called the (four-) velocity and mv is called the (four-) momentum.

Light consists of photons which are particles of mass zero travelling on null geodesics. Along a null geodesics, there is no length, but a parameter is well determined up to affine transformation (i. e. , replacing t by at + b where a and b are constants). A photon has a momentum, which amounts to a special choice of the possible parametrizations

of the null geodesic. This gives a well defined momentum
vector

$$p = \left(\frac{dx^1}{ds}, \cdots, \frac{dx^4}{ds} \right) .$$

Having regarded (7.3) as the appropriate generali-
zation of Newton's laws for motion in the presence of a
gravitational field, we must now find the equations determining the field itself. It should specialize to the Laplace
equation $\Delta \varphi = 0$ in vacuo and the Poisson equation $\Delta \varphi = \rho$
in the presence of matter. For this purpose we recall the
definition of the various curvatures of a Riemanian (or
Lorentzian) manifold. The Riemann curvature tensor is
defined by

$$R(X, Y) = [\nabla_X, \nabla_Y] - \nabla_{[X, Y]}$$

where $R(X, Y)$ is a linear transformation of vector fields
(depending on X and Y) i.e., $Z \rightarrow R(X, Y)Z$ is given by
applying the right hand side to Z . Letting $\delta_i = \frac{\partial}{\partial x^i}$ in a
given coordinate system we have

$$R(\delta_i, \delta_j)\delta_k = \nabla_{\delta_i}\left(\nabla_{\delta_j}\delta_k\right) - \nabla_{\delta_j}\left(\nabla_{\delta_i}\delta_k\right) \qquad \text{since } [\delta_i, \delta_j] = 0$$

$$= \nabla_{\delta_i}\left(\sum_\ell \Gamma^\ell_{jk}\delta_\ell\right) - \nabla_{\delta_j}\left(\sum_\ell \Gamma^\ell_{ik}\delta_\ell\right)$$

$$= \sum \left\{ \frac{\partial \Gamma^\ell_{jk}}{\partial x^i} - \frac{\partial \Gamma^\ell_{ik}}{\partial x^j} + \Gamma^m_{jk}\Gamma^\ell_{im} - \Gamma^m_{ik}\Gamma^\ell_{jm} \right\}\delta_\ell .$$

The coefficients of δ_ℓ is denoted by R^ℓ_{kij} so that

$$R(\delta_i, \delta_j)\delta_k = \sum R^\ell_{kij}\delta_\ell$$

where

$$(7.4) \qquad R^\ell_{kij} = \frac{\partial \Gamma^\ell_{ik}}{\partial x^i} - \frac{\partial \Gamma^\ell_{ik}}{\partial x^j} + \Gamma^m_{jk}\Gamma^\ell_{im} - \Gamma^m_{ik}\Gamma^\ell_{jm} .$$

The Ricci curvature tensor is defined as: if X, Z are vector fields on M, then

$Ric(X, Z) = $ Trace of the linear transformation:

$Y \rightsquigarrow - R(X, Y)Z$

(Trace is taken pointwise on M). It depends on X and Z and is easily seen that

$$(7.5) \qquad Ric(\delta_i, \delta_k) \equiv R_{ik} = -\sum_\ell R^\ell_{ik\ell} .$$

The scalar curvature is defined to be the trace of t

the linear transformation associated to the Ricci curvature

and is given, in local coordinates, by

(7. 6) $$R = \Sigma\, R_{ii}\ .$$

Using the same coordinates as introduced above (that is,

$g_{4i} = 0$, $i = 1, 2, 3$ and we ignore quadratic terms in the

Γ's)

$$R_{44} = -\sum_{\alpha} \left(\frac{\partial \Gamma^{\alpha}_{4\alpha}}{\partial x^4} - \frac{\partial \Gamma^{\alpha}_{44}}{\partial x^{\alpha}} \right)$$

$$= -\frac{1}{2}\sum \left\{ \frac{\partial^2 g_{44}}{\partial x^{\alpha\,2}} - \frac{\partial^2 g_{\alpha\alpha}}{\partial x_4^{2}} \right\}\ .$$

Now if we are close enough to the central geodesic we

would expect the second term in the above expression to be

negligible in comparison with the first. (Also, the observed

gravitational field appears to be stationary.) Thus $R_{44} =$

$-\frac{1}{2}\Delta g_{44}$ so that $R_{44} = 0$ gives Laplace's equation. In

order to make this invariant Einstein substitutes the

equation

(7. 7) $$G_{ij} = R_{ij} - \frac{1}{2}\, g_{ij} R = 0$$

in the absence of matter (or energy). (Actually in the
presence of matter-energy the Einstein field equations
read

$$G_{ij} - \Lambda G \, g_{ij} = k T_{ij}$$

where T_{ij} is the so-called energy momentum tensor and
Λ and k are constants. In "empty" space time $T_{ij} = 0$
and Λ is so extremely small it can be neglected so we
obtain (7.7).)

Let us now assume that we are dealing a spherically
symmetric field. For instance, suppose that the central
geodesic of the coordinate system represents the history of
the sun (idealized to a particle). Suppose that the metric
is such that any rotation of the normal bundle expontiates
(locally) to an isometry of space time. There is a unique
solution of (7.7) satisfying this symmetry condition. It is
found as follows: We first introduce a special coordinate
system. Let τ denote the coordinate along the geodeisc
and ρ the distance to the geodesic so that r, θ, ψ are
polar coordinates in the fiber of the normal bundle at τ .
By assumptions, the surface $(r, \tau) = \text{const}$ is a two -
dimensional sphere, (since it is a compact two dimension
surface of constant curvature by the homogeneity assumption).

The metric on this sphere is $-r^2(d\theta^2 + \sin^2\theta d\varphi^2)$ if θ, φ are polar coordinates and r is the radius. Consider the family of three dimensional surfaces given by $r = \mathrm{const}$. The orthogonal trajectory to these surfaces passing through (r_0, θ_0, ψ_0) will pass through the

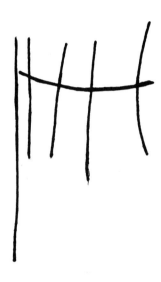

geodesic (and intersect it orthogonally). By spherical symmetry these curves will fit together to form three dimensional surfaces; that is, the parameter of the point, $t(r, \theta, \psi)$,of intersection of this trajectory with the geodeisc gives a well defined coordinate. We use r, θ, φ , and t as local coordinates excluding $r = 0$. Then, in view of the orthogonality, (and Gauss's lemma [S 13] p. 201) the metric takes the form

(7.8) $-A(r,t)dr^2 - r^2(d\theta^2 + \sin^2\theta^2 d\varphi^2) + B(r,t)dt^2$.

One now computes the curvature tensor and the Ricci curvature tensor [S 14] p. 272-274. This is a tedious but straightforward computation which we will not reproduce here. The end result is that the Einstein equation (7.7) imply that the functions A and B are independent of t (after a suitable change in the time scale t). This is known as Birkhoff's theorem. We also find that, for $r > 2M > 0$ (where M is a constant to be identified with the mass of the sun)

(7.9)

$$A = \frac{1}{1 - \dfrac{2M}{r}}$$

$$B = \left(1 - \frac{2M}{r}\right) .$$

This is known as the Schwartzchild solution so that (7.8) with these values of A and B is called the Schwartzchild metric.

Let us write the equations of geodesics in a Lagrangian form (cf. [L 3]). The Lagrangian corresponding to the metric $\Sigma_{i,j} g_{ij} dx^i dx^j$ has the form

$$L = \frac{1}{2} \sum_{i,j} g_{ij} \dot{x}^i \dot{x}^j$$

where $\dot{x}^i = \frac{dx^i}{ds}$, with s = "arc length" parameter. A geodesic is given by the equations

$$\frac{\partial L}{\partial x^i} - \frac{d}{ds}\left(\frac{\partial L}{\partial \dot{x}^i}\right) = 0 \quad .$$

In the case of the Schwartzchild metric, it is immediately seen that the Lagrangian does not contain t or φ. Thus from the equations for t and φ, we see that

$$\frac{\partial L}{\partial \dot{t}} = \left(1 - \frac{2M}{r}\right)\dot{t}$$

and

$$\frac{\partial L}{\partial \dot{\varphi}} = r^2 \sin^2 \theta \cdot \dot{\varphi}$$

are constants of motion.

The equation for θ becomes

$$\frac{d}{ds}(r^2 \dot{\theta}) - r^2 \sin \theta \cos \theta \, \dot{\varphi}^2 = 0 \quad .$$

Thus if $\theta(0) = \frac{\pi}{2}$ and $\dot{\theta}(0) = 0$ then $\theta \equiv \frac{\pi}{2}$ is a solution. Since, for a particular geodesic, we can always rotate coordinates to obtain these initial conditions, we see that there is no loss of generality in restricting attention to the "plane" $\theta = \frac{\pi}{2}$.

Let us first examine light rays, i.e., null geodesics. Along such a geodesic we have

$$(7.10) \qquad \left(\frac{1}{1 - \frac{2M}{r}}\right) dr^2 + r^2 d\varphi^2 = \left(1 - \frac{2M}{r}\right) dt^2 \quad .$$

Let us regard

$$\left(\frac{dr}{dt}\right)^2 + r^2\left(\frac{d\varphi}{dt}\right)^2 = v^2$$

as the "velocity of light as given by the local coordinates". That is, let us introduce the Euclidean metric

$$d\sigma^2 = dr^2 + r^2 d\varphi^2 \quad .$$

Then

$$\left[\frac{1}{\left(1 - \frac{2M}{r}\right)} \left(\frac{dr}{d\sigma}\right)^2 + r^2 \left(\frac{d\theta}{d\sigma}\right)^2\right] v^2 = \left(1 - \frac{2M}{r}\right) \quad ,$$

or setting

$$\frac{dr}{d\sigma} = \cos\eta$$

$$r\frac{d\theta}{d\sigma} = \sin\eta$$

$$(7.11) \qquad v^2 = \frac{\left(1 - \frac{2M}{r}\right)^2}{\cos^2\eta + \sin^2\eta\left(1 - \frac{2M}{r}\right)} \qquad .$$

We thus see that for r not much greater than 2M we have that v is less than one by a non-negligible amount. In particular, for a radial direction $(\eta = 0)$

$$v = \left(1 - \frac{2M}{r}\right)$$

and for a transverse direction $(\eta = \frac{\pi}{2})$

$$v = \left(1 - \frac{2M}{r}\right)^{1/2} \qquad .$$

In any event, light "slows down" near a heavy body. It is exactly this effect that has recently been measured in the radar experiment of Shapiro, Ash et. al, at Linclon Laboratories. After the orbit of Mercury is carefully computed by numerical integration (with constant correct-ion by least square methods for the appropriate constants and initial conditions) a radar beam is sent to Mercury at superior conjunction. Then the delay in receiving the echo is carefully measured. This then determines whether or not light slows down near the sun. As of the present writing (May 1968) the evidence seems to strongly support, in a quantitative way, the general theory of relativity.

Since the radial and transversal velocities are diminished we should expect that the geodesics will be "curved" in the neighborhood of the sun. This is indeed the case, cf.[S14]p. 297. This was quantitatively verified by observing the stars near the sun during a solar eclipse. However this effect has not provided a quantitatively satisfactory verification. The most famous prediction of general relativity is the advance of the perihelion of Mercury:

Let us consider particle of mass 1 (the planet) moving along a timelike geodesic of (7.8)-(7.9) . Then as above we have the integrals

$$r^2\dot{\varphi} = G = const$$

$$\left(1 - \frac{2M}{r}\right)\dot{t} = k = const$$

where $\cdot$ denotes differentiation with respect to s . Also, since s is the time parameter

$$\left(1 - \frac{2M}{r}\right)\dot{t}^2 - \left(1 - \frac{2M}{r}\right)^{-1}\dot{r}2 - r^2\dot{\varphi}^2 = 1 \quad .$$

To obtain the equation of the orbits we write $\dot{r} = \frac{dr}{d\varphi}\dot{\varphi}$ to obtain

$$\frac{k^2}{\left(1 - \frac{2M}{r}\right)} - \frac{1}{\left(1 - \frac{2M}{r}\right)} \frac{G^2}{r^4} \left(\frac{dr}{d\varphi}\right)^2 - \frac{G^2}{r^2} = 1$$

or

$$\frac{G^2}{r^4} \left(\frac{dr}{d\varphi}\right)^2 + \left(1 + \frac{G^2}{r^2}\right)\left(1 - \frac{2M}{r}\right) = k^2 \quad .$$

Setting

$$z = \frac{1}{r}$$

this becomes

$$\left(\frac{dz}{d\varphi}\right)^2 = \frac{k^2 - 1}{G^2} + \frac{2M}{G^2} z - z^2 + 2Mz^3$$

or (if we assume non-negligible eccentricity so that $\frac{dz}{d\varphi}$ has only isolated zeros) we get, by differentiating,

$$\frac{d^2 z}{d\varphi^2} + z = \frac{M}{G^2} + 3Mz^2 \quad .$$

If we could ignore the term $3Mz^2$ we would obtain Kepler motion

(7.12) $\qquad \dfrac{\overline{1}}{r} = \overline{z} = \dfrac{M}{G^2} \left[1 + e \cos (\varphi - g)\right] \quad .$

The exact solution of the equation is expressible in terms

of elliptic integrals and can be found in [S14] p. 293. We shall proceed approximately as follows: Since the elliptical motion is a good approximation, if we substitute $\bar{z}^2$ for z^2 in our equation we will be making a quadratic error. Our approximate equation reads

$$\frac{d^2 z}{d\varphi^2} + z = \frac{M}{G^2} + 3M\bar{z}(\varphi)^2 \ .$$

This is now a linear differential equation. The linear equation

$$\frac{d^2 w}{d\varphi^2} + w = 3M\bar{z}(\varphi)^2$$

$$= \frac{3M^3}{G^4} \left\{ 1 + 2e\cos(\varphi - g) + \frac{1}{2}e^2[1 + \cos 2(\varphi - g)] \right\}$$

has the explicit solution

$$w = \frac{3M^3}{G^4} \left\{ 1 + \frac{1}{2}e^2 + e\,\varphi\sin(\varphi - g) - \frac{1}{6}e^2\cos 2(\varphi - g) \right\} \ .$$

This is to be added to $\bar{z}(\varphi)$. The first of these terms consists of adding a small constant value to $\bar{z}$ and can be ignored, as can the third, which is small and periodic. Thus, approximately

$$z = \frac{M}{G^2}[1 + e\cos(\varphi - g)] + \frac{3M^3}{G^4}e\varphi\sin(\varphi - g) \ ,$$

which equals,

$$\frac{M}{G^2}\left[1 + e\cos\left(\varphi - g - \frac{3M^2}{G^2}\varphi\right)\right] + \cdots$$

where we use the Taylor expansion for $\cos\theta$ and ignore terms which are quadratic in $\frac{3M^2}{G^2}$. (M is a small quantity in the units we are using.) This then gives a slow advance of the perihelion of Mercury. Unfortunately, it is at present open to debate whether this is confirmed by observation. There is an observed advance of Mercury's perihelion of about 5599 seconds of arc per century. Classical perturbation theory predicts an advance of 5557". The above formula predicts 43" . Thus if we brutally add the Einstein correction to the classical Newtonian perturbation we would get the observed advance. Unfortunately, it is open to debate whether the various constants (such as the planetary masses, the mass distribution of the sun etc) are known to such accuracy so as to justify saying that a correction of 1% is evidence of the correctness of general relativity.

Hundreds of thousands of optical and radar observations are now being processed at Lincoln Laboratories (by Ash, Shapiro, et. al) and it is to be hoped that something definitive can be said on this point in the near future.

148.

REFERENCES

[A 1] ADAMS, J. C., "Lectures on the Lunar Theory"
Collected Scientific Papers, vol. ll, pp. 1-84.
    One of the classical expositions of the lunar theory.
It is difficult going for the modern reader due to
notational problems. I prefer the exposition in
Darwin [D 2] .

[A 2] ANOSOV, D.B., "Geodesic flow on compact
Riemannian manifolds of negative curvature" (in
Russian) TRUDY. MAT. INST. STELOVA. 1967.

[A 3] ARENTSORF, R. F., "Periodic solutions of the
restricted three body problem representing analytic
continuation of Keplerian elliptic motion." Amer.
Jour. of Math. vol. 85 (1963) pp. 27-

[A 4] ARNOLD, V. I., "Small denominators, I, on the
mapping of a circle into itself". Izvestia Akad.
Nauk. serie Math. 25, 1 (1961) p. 21-86 = Trans.
Amer. Math. Soc. 2 series vol. 46, p. 213-284.

[A 5] _____ , Proof of a theorem of A. N. Kolmogorov
on the invariance of quasi-periodic motions under
small perturbations of the Hamiltonian. Uspehi
Math. Nauk., vol. 18, ns (1963), p. 13-40 = Russian
Math Surveys, vol. 18, no. 5, p. 9-36.

149.

[A 6] _____ , Small denominators and problems of stability of motion in classical and celestial mechanics. Uspehi Math. Nauk., vol. 18, no. 6 (1963), p. 91-196 = Russian Math Surveys, vol. 18, no. 8, p. 85-193 .

[A 7] ARNOLD, V.I. and AVEZ, A., Problèmes ergodiques de la mecanique classique. Gauthier-Villars, Paris, 1967.

[B 1] BARRAR, R.B., "Existence of periodic orbits of the second kind in the restricted three body problem", Astron. Jour. 70 (1965), pp. 3-6.

[B 2] BIRKHOFF, G.D., Dynamical Systems, American Math. Society Colloquium Publication, vol. IX, New York (1927).

[B 3] _____ , "The restricted problem of three bodies bodies". Rendiconti Palermo, vol. 39 (1915) pp. 1-70 Collected Papers vol. 1, pp. 682-751.

[B 4] _____ , "Sur le problème restreinte des trois corps", Annali della Scuola Normale Superiore de Pisa 1935, s. 2, vol. 4, pp. 267-306. Collected Papers, vol. 2, pp. 466-505.

[B 5] _____ , "Proof of Poincaré's geometric theorem", Trans. Amer. Math. Soc., vol. 14 (1913), pp. 14-22.

[B 6] _____ , "An extension of Poincaré's last geometric theorem", Acta Math. , vol. 47, (1925), pp. 297-311, Collected Papers, vol. II, pp. 252-266.

150.

[B 7] _____ , "Surface transformations and their dynamical applications", Acta Mathematica, vol. 43, 1920, pp. 1-119, Collected Math. Papers, pp. 111-230.

[B 8] _____ , "Proof of the ergodic theorem", Proc. Nat. Acad. Sci., vol. 17 (1931), pp. 636-660. Coll. Math. Papers, vol. 2, pp. 404-408.

[B 9] _____ , and D. C. LEWIS, JR., "On the periodic motions near a given periodic motion of a dynamical system.", Ann. di. Math., s. 4, vol. 12, pp. 117-133, Collected Math Papers, vol. 2, pp. 820-836.

[B 10] BOHL, P., "Uber un in der Theorie der säkuloren Stoungen vorkommedes Problem", Jour. reine und ang.   Math. (2), vol. 135 (1909), pp. 189-283.

[B 11] _____ , "Bemerkungen zur Theorie der säkularen Störungen". Math. Ann. vol. 72, (1912) pp. 295-296.

[B 12] BOHR, N, "Zur Theorie der fastperiodeschen Functiones" I, II, III, Acta Math., vol. 45 (1924) pp. 29-127, vol. 46(1925) pp. 101-214 vol. 47 (1926) pp. 237-281 .

[B 13] D. BROUWER and G. M. CLEMENCE, Methods of Celestial Mechanics. Academic Press, New York (1961).

[B 14] BROWN, E. W., An introductory treatise on the lunar theory, Cambridge University Press, 1896.

[B 15] _____ , Article on the moon's path in the
Encyclopedia Americana, 1949 Edition.

[C 1] CARTAN, E. , Leçons sur les Invariants intégraux,
Hermann and Co. , Paris 1922.

[C 2] CHARLIER, C. V. L. , Die Mechanik des Himmels,
Two volumes, W. de Gruyter and Co. , Leipzig
(192 7).

[C 3] CHAZY, J. , La theorie de la relativité et la
mecanique celeste. Two volumes, Gauthier-
Villars, Paris 1928 and 1930.

[D 1] DANBY, J. M. A. , Fundamentals of celestial
mechanics. The Macmillan Co. , New York, 1962.

[D 2] DARWIN, Sir G. H. , "Lectures on the lunar theory",
Collected Works, Volume 5 .

[D 3] DENJOY, A. , "Sur les courbes defines par les
équations différentielles à la surface du tore", J.
Math. Pures et Appls. vol. 11 (1932) pp. 333-375 .

[E 1] EINSTEIN, A. , The meaning of relativity,
Princeton University Press.

[F 1] FAVARD, J. , Lecons sur les fonctions presques
périodiques. Paris 1933.

[F 2] FERMI , E , "Beweiss, dass ein mechanisches
Normalsystem im allermein quasi-ergodisch ist"
Z. Physik, vol. 24 (1923) p. 261.

152.

[F 3] FINZI, A., "Sur le problème de la génération d'une
transformation donnée d'une courbe fermée par une
transformation infinitesimal", Ann. Sci. Ecole.
Norm. Sup., vol. 6 7 (1950) pp. 273-305; vol. 69
(1952) pp. 3 71-430.

[G 1] GANDZ, S., Translation of the code of Maimonides.
Book three treatise eight-Sanctification of the New
Moon. Yale Judaica Series Vol. XI, Yale University
Press, New Haven, Conn. 1956.

[G 2] GARSIA, A. M., "A simple proof of E. Hopf's
maximal ergodic theorem", Jour. Math. and Mechs.
vol. 14, no. 3 (1965).

[G 3] GOLDSTEIN, H., Classical Mechanics, Addison-
Wesley Publishing Co., Reading, Mass. 1950.

[H 1] HADAMAND, J., "Sur l'iteration et les solutions
asymptotiques des equations differentielles", Bull.
Soc. Math. France, vol. 24 (1901)pp. 224-5.

[H 2] HARTMAN, P., Ordinary differential equations,
Wiley, New York, 1964.

[H 3] _____, "Mean motions and almost periodic
functions". Trans. Amer. Math. Soc., vol. 49
(1939) pp. 64 - 81.

[H 4] HILL, G. W., "Researches in the lunar theory",
Amer. Journal of Math. vol. 1 (1878), pp. 5, 129,
245. Collected Math Works, vol. 1, pp. 284.

[J 1]   JESSEN, B. and TORNEHAVE ,H. "Mean motions
        and almost periodic functions", Acta Mathematica,
        vol. 77 (1945) pp. 137-279.

[K 1]   KOESTLER, A. , The Sleepwalkers, The Macmillan
        Co. , New York, 1959.

[K 2]   KOLMOGOROV, A. N. , "General theory of
        dynamical systems and classical mechanics", Proc.
        Int. Cong. Math. , North Holland, Amsterdam
        (1957) pp. 287 -331 .

[L 1]   LA GRANGE, J. L. , "Théorie des variations
        séculaires des éléments des planetes". Oeuvres
        vol. 5, pp. 123-344.

[L 2]   LANCZOS, C. , The variational principles of
        mechanics. University of Toronto Press, Toronto,
        1949.

[L 3]   LOOMIS, L. , and STERNBERG, S. , Advanced
        Calculus. Addison-Wesley, Reading, Mass. , 1968.

[M 1]   MOSER, J. , "On the integrability of area preserving
        Cremona mappings near an elliptic fixed point. "
        Bull. Soc. Mat. Mexicana, vol. 2, no. 5 (1960),
        pp. 1 76-180.

[M 2]   _____ , "On invariant curves of area preserving
        mappings of an annulus. " Nach. Akad. Wiss. Gott.
        Math. Klass (1962), pp. 1-20.

[M 3]   _____ , "Some aspects of non-linear differential
        equations. ", Ann. Scuola. Norm. Sup. Pisa. 1966.

154.

[M 4] _____ , "Convergent series expansions for
quasi-periodic motions". Math. Ann. , vol. 169
(1967) pp. 136-176.

[N 1] NASH, J. , "The embedding problem for Riemann
manifolds.", Ann. of Math., vol. 63(1956)pp. 20-63.

[N 2] NEUGEBAUER, O. , The exact sciences in
antiquity. Brown University Press, Providence,
Rhode Island 1957.

[N 3] _____ , Astronomical appendix to the translation of Maimonides code, cf. [G 1] .

[N 4] von NEUMANN, "Almost periodic functions in a
group", Trans. Amer. Math. Soc. , vol. 36 (1934)
pp. 445-492.

[P 1] PANNEKOEK, A. E. , A History of Astronomy,
George Allen and Unwin Ltd. , London, 1961.

[P 2] POINCARÉ , H. , "Sur les courbes definies par
les équations differentielles", J. Math. Pures Appl.
s. 3, vol. 7 (1881) 375-422; s. 3, vol. 8 (1882), pp.
251-286, s. 4, vol. 1 (1885) pp. 167-244, s. 4, vol. 2
(1886) pp. 151-217.

[P 3] _____ , "Sur les déterminants d'ordre infini",
Bull. Soc. Math. France, vol. 14 (1886), pp. 76

[P 4] _____ , "Sur le déterminant de Hill", Bull.
Astron. vol. 17 (1900), pp. 134.

[P 5] _____ , Les Methodes Nouvelles de la
Mécanique Celeste, Three volumes, Gauthier-
Villars, Paris 18

[P 6] _____ , Leçons de Mécanique Celeste, Three
volumes, Gauthier-Villars, Paris, 1905-10.

[P 7] _____ , Cours d'Astronomre général
(Polycopié) Ecole Polytechinique, Paris (undated).

[P 8] POLLARD, H. , Mathematical introduction to
celestial mechanics, Prentice-Hall, Inc, Englewood
Cliffs, N. J. , 1966.

[R 1] ROKHLIN, V. A. , "Exact endomorphisms of
Lebesque spaces", Izv. Math. Nauk. , vol. 25
(1961) pp. 499-530, Amer. Math. Soc. Translation
series 2, vol. 39 (1964) pp. 1-36 .

[S 1] SCHWARTZ, J. , Non-linear analysis, Lecture
notes, New York University (19 64) .

[S 2] SCHWARTZMAN, S. , "Asymptotic cycles", Annals
of Math. , vol. 66 (195 7) pp. 270-284.

[S 3] SIEGEL, C. L. , Vorlesungen uber Himmelsmechanik.
Springer, Berlin (1956).

[S 4] _____ , "Note on differential equations on the
torus", Ann. of Math. vol. 46 (1945) pp. 423 - 428.

[S 5] "Uber die Normalform analytischer Differential-
gleichungen un die Nahe einer Gleichgewichtlosung. "
Nach. Akad. Wiss. Gott (1952), pp. 21-30.

156.

[S 6] _____ , "Uber die Existece einer Normalform
       analytischer Hamiltonische Differentialglichungen
       in der Nahe einer Gleichgewichtlosung." Math.
       Ann. vol. 28 (1954) pp. 144-170.

[S 7] SMALE, S. , "Differentiable dynamical systems.",
       Bull. Amer. Math. Soc. , vol. 73 (1967) pp. 747 -
       817.

[S 8] SMART, W. M. , Celestial Mechanics, Longmanns,
       Green, London 1953.

[S 9] STERNBERG, S. , "On differential equations on the
       torus", Amer. J. Math. vol. 79 (195 7)

[S 10] _____ , "Local contractions and a theorem
        of Poincaré ", Amer. J. Math. , vol. 79 (1957)
        pp. 809-824.

[S 11] _____ , "The structure of local homeomorph-
        isms III", Amer. J. Math. , vol. 81 (1959) pp. 578-
        604.

[S 12] _____ , "Infinite Lie groups and the formal
        aspects of dynamical systems", J. Math. Mech. ,
        vol. 10 (1961) pp. 451-474 .

[S 13] _____ , Lectures on differential geometry,
        Prentice-Hall, Inc. , Englewood Cliffs, N. J. 1964.

[S 14] SYNGE, J. L. , Relativity, the general theory,
        North Holland Publishing Co. , Amsterdam, 1960.

[T 1] TISSERAND, F. , Traité de mécanique céleste .
       Four volumes, Gauthier-Villars, Paris .

[W 1] WATSON, G. N. , Treatise on the Bessel functions, Cambridge University Press (1958), pp. 411-420.

[W 2] WEIL, A. , L'integration dars les groupes topologiques. Hermann and Co. , Paris, 1953 .

[W 3] WEYL, H. , Space, time, matter. Dover Publishing Co. , New York.

[W 4] _____ , "Sur une application de la theorie des nombres a la mechanique statistique et la theorie des perturbations", Enseign. Math. vol. 16 (1914) pp. 455-467.

[W 5] _____ , "Uber die gleichverteilung mod eins. " Math. Ann. , vol. 77 (1916) pp. 313-352.

[W 6] _____ , "Mean motion I, II, Amer. J. of Math. vol. 60 (1938) pp. 889-896, vol. 61 (1939) pp. 143-148.

[W 7] WHITTAKER, E. T. , A treatise on the Analytical dynamics of Particles and Rigid Bodies. Cambridge University Press, London, 1937.

[W 8] _____ , and ROBINSON, G. , The calculus of observations, Blacke and Son Ltd. , 1924.

[W 9] WINTNER, A. , Analytical foundations of celestial mechanics. Princeton University Press, 1941.

[W 10] _____ , Zur Hillschen Theorie der Variation des Mondes. Math. Zeit. , vol. 24, (1926), pp. 259-265.

158.

[Y 1]   YOSHIDA, K. , <u>Functional Analysis</u>. Springer-
Verlag, Berlin, 1965.